MIX
Papier aus verantwortungsvollen Quellen
Paper from responsible sources
FSC® C105338

**Dr. Bikram Jit Singh
Dr. Dinesh Khanduja**

WRAP THE SCRAP WITH DMAIC

Strategic Deployment of Six Sigma in Indian Foundry SMEs

Anchor Academic
Publishing

Dr. Singh, Bikram Jit. Dr. Khanduja, Dinesh: WRAP THE SCRAP WITH DMAIC: Strategic
Deployment of Six Sigma in Indian Foundry SMEs, Hamburg, Anchor Academic
Publishing 2015

Buch-ISBN: 978-3-95489-395-9
PDF-eBook-ISBN: 978-3-95489-880-0
Druck/Herstellung: Anchor Academic Publishing, Hamburg, 2015

Bibliografische Information der Deutschen Nationalbibliothek:
Die Deutsche Nationalbibliothek verzeichnet diese Publikation in der Deutschen
Nationalbibliografie; detaillierte bibliografische Daten sind im Internet über
http://dnb.d-nb.de abrufbar.

Bibliographical Information of the German National Library:
The German National Library lists this publication in the German National Bibliography.
Detailed bibliographic data can be found at: http://dnb.d-nb.de

All rights reserved. This publication may not be reproduced, stored in a retrieval system
or transmitted, in any form or by any means, electronic, mechanical, photocopying,
recording or otherwise, without the prior permission of the publishers.

Das Werk einschließlich aller seiner Teile ist urheberrechtlich geschützt. Jede Verwertung
außerhalb der Grenzen des Urheberrechtsgesetzes ist ohne Zustimmung des Verlages
unzulässig und strafbar. Dies gilt insbesondere für Vervielfältigungen, Übersetzungen,
Mikroverfilmungen und die Einspeicherung und Bearbeitung in elektronischen Systemen.

Die Wiedergabe von Gebrauchsnamen, Handelsnamen, Warenbezeichnungen usw. in
diesem Werk berechtigt auch ohne besondere Kennzeichnung nicht zu der Annahme,
dass solche Namen im Sinne der Warenzeichen- und Markenschutz-Gesetzgebung als frei
zu betrachten wären und daher von jedermann benutzt werden dürften.

Die Informationen in diesem Werk wurden mit Sorgfalt erarbeitet. Dennoch können
Fehler nicht vollständig ausgeschlossen werden und die Diplomica Verlag GmbH, die
Autoren oder Übersetzer übernehmen keine juristische Verantwortung oder irgendeine
Haftung für evtl. verbliebene fehlerhafte Angaben und deren Folgen.

Alle Rechte vorbehalten

© Anchor Academic Publishing, Imprint der Diplomica Verlag GmbH
Hermannstal 119k, 22119 Hamburg
http://www.diplomica-verlag.de, Hamburg 2015
Printed in Germany

DEDICATED TO OUR BELOVED FATHERS

PREFACE

In present age of economic turbulence, Indian foundries are striving hard to achieve overall operational excellence to remain competitive. To cope up with the emerging future challenges, 'Defects Reduction' leading to less scrap seems to be the most promising and viable strategy to achieve higher productivity in foundries. This can be remarkably tackled by inculcating Six Sigma's DMAIC approach in the given production environments as Six Sigma is evolving into a powerful business improvement strategy and its importance is growing every day. For achieving and sustaining operational excellence, Six Sigma enables companies to use simple but powerful statistical methods to define, measure, analyse, improve and control processes Doing things rightly and keeping them consistent is the basic fundamental idea behind Six Sigma. The literature review suggests that in India, except in some multinational companies, the application of Six Sigma has been highly inconsistent in manufacturing sector and is almost non-existent in the foundry sector which is facing serious sickness levels.

Literature also reports tremendous financial gains and productivity improvement through Six Sigma in large manufacturing sector, so an attempt can be made to replicate it in foundry SME sector too. Keeping this in mind, this book was planned to explore the synergy of Six Sigma and foundry industry. Through a case study in a non ferrous medium scale foundry, the present work formulates a comprehensive strategy for the successful implementation of Six Sigma in Indian foundry industry without ignoring its existing constraints. Using a project based approach; standard step by step methodology has been formulated to execute different phases of DMAIC. Selecting right tools/techniques in various DMAIC phases has always been a critical parameter for overall success of the Six Sigma project in the presence of some real constraints/difficulties. In this context, a new framework has been suggested for executing each phase, which helps in right selection of tools/techniques. This book is an attempt to dispel different myths prevailing regarding Six Sigma among Indian SMEs, specifically in the foundry units. An extensive use of various tools has been made during various phases and these include:

Define Phase (Voice of Customer, QFD, Project Charter, Project Scheduler, Snaps of Problem, Historical analysis of problem, COPQ Matrix, Process by SIPOC Diagram and Project Goals);

Measure Phase (Sigma Calculator, Cpk study, Pareto Charts, FMEA Table, Cause and Effect Matrix, Gauge R&R study, Bias Checking and Stability Test);

Analyze Phase (Chi Square Test, One Way ANOVA, Two Sample t-Test, Multi Regression, Interaction Plot, Fish Bone Diagram, Why Why Analysis);

Improve Phase (Design of Experiments (DOE), Poka-Yoke and Kaizen) and

Control Phase (Control Plan, X bar and R Chart for BT variation and p-Chart for overall scrap tracking).

      By reducing the scrap of pistons from around 22% to 10%, improvement in the sigma level from 3.43 to 3.67 have been achieved by implementing the suggested approach. Even this small increase has resulted in savings of Rs. 30.7 lakhs per annum and this is remarkable for a medium scale make-to-order foundry unit. Inferences from the findings have a significant impact on the foundry sector. This study has vitally reaffirmed the efficacy of Six Sigma strategy in Indian foundry industry to reduce scrap/waste from the operations, thus greatly improving the production efficiency. Researchers have laid more emphasis on defining various tools and techniques of Six Sigma but 'tool selection criterion' for each phase, as per the given constraints and environment, is not available to industries. Due to lack of this vital information, usually wrong tools and techniques are chosen and so DMAIC project fails by moving in a wrong direction, with course of time. A cadre with sound theoretical knowledge on different statistical tools and software needs to be built up in the management, so as to bridge the gap between the theory and practice of Six Sigma and appreciate its potential while bringing in business excellence. The challenge for all organizations is to integrate Six Sigma into their core business processes and operations rather than managing it as a separate initiative.

ACKNOWLEDGEMENT

This acknowledgement is intended to be thanks giving gesture to all those people who have been involved to complete this work directly or indirectly.

First and foremost, we would like to express our thanks and indebtedness to Er. Maninderpal Singh (Master Black Belt, Gabriel India Ltd. Parwanoo) to provide expert advice whenever needed. We are highly obliged to him for being there always whenever we needed him.

We would like to express our sincere gratitude to Dr. Sudhir Saxena (Professor and HOD), Department of Mechanical Engineering, NIT, Kurukshetra, for his unfailing support and constant guidance throughout all stages of this book.

We express our deep sense of gratitude and sincere thanks to the staff of Federal Mogul Goetze India Limited Bahadurgarh, Patiala for their support and providing access to data/documents/processes needed during the project. We wish to extend our sincere thanks to Er. Daljeet Singh (Head, Piston Foundry, Federal Mogul India Ltd. Patiala) for his benign help and cooperation throughout the project work (as discussed in the given book).

Finally, we would like to appreciate our parents, who have always been a great source of support and encouragement, especially in all of our academic endeavours.

Dr. BIKRAM JIT SINGH
&
Dr. DINESH KHANDUJA

CONTENTS

	PAGE NO.
Preface	iii-iv
Acknowledgement	v
Contents	vi-vii
List of Figures	viii-x
List of Tables	xi
Nomenclature	xii-xiii

CHAPTER 1:		**INTRODUCTION**	**01-17**
	1.1	Pretext	01
	1.2	Quality Control in Foundries through Six Sigma	02
	1.3	Six Sigma: Origin and Development	03
	1.4	Six Sigma: A Statistical Concept	06
	1.5	Implementation of Six Sigma	09
	1.6	Motivation for the Present Book	15
	1.7	Objectives of the Book	17

CHAPTER 2:		**ANTECEDENTS OF SIX SIGMA**	**18-38**
	2.1	Evolution of Six Sigma	18
		2.1.1 Definition of Six Sigma	18
		2.1.2 Major Themes Emphasized	20
		2.1.3 Six Sigma Tools and Techniques	23
	2.2	Six Sigma in Manufacturing Sector	24
	2.3	Six Sigma in Indian Industries	26
	2.4	Six Sigma in Foundry Sector	30
	2.5	Summary	37

CHAPTER 3:		**RESEARCH DESIGN**	**39-54**
	3.1	Problem Formulation	39
	3.2	Book Plan	39
	3.3	Methodology Adopted	40
	3.4	Tools Used in Present Case Study	44

CHAPTER 4:		**AN INDIAN CASE STUDY**	**55-137**
	4.1	Introduction	55
	4.2	About the Organization	55
	4.3	Implementation of DMAIC Approach	57
	4.4	Define Phase	58
		4.4.1 Proposed Framework for D-Phase	61
		4.4.2 Case Findings (D-Phase)	63

		4.4.3 Inferences from D-Phase	69
	4.5	Measure Phase	69
		4.5.1 Proposed Framework for M-Phase	71
		4.5.2 Case Findings (M-Phase)	72
		4.5.3 Inferences from M-Phase	84
	4.6	Analyse Phase	84
		4.6.1 Proposed Framework for A-Phase	85
		4.6.2 Case Findings (A-Phase)	88
		4.6.3 Inferences from A-Phase	101
	4.7	Improve Phase	102
		4.7.1 Proposed Framework for I-Phase	103
		4.7.2 Case Findings (I-Phase)	106
		4.7.3 Inferences from I-Phase	124
	4.8	Control Phase	125
		4.8.1 Proposed Framework for C-Phase	126
		4.8.2 Case Findings (C-Phase)	128
		4.8.3 Inferences from C-Phase	134
	4.9	Result Appraisal	134

CHAPTER 5:   IMPLICATION AND SCOPE   138-142

5.1	Conclusions	137
5.2	Scope	141

BIBLOGRAPHY   **143-170**

ANNEXURE-1	171-173
ANNEXURE-2	174
ANNEXURE-3	175-176
ANNEXURE-4	177
ANNEXURE-5	178
ANNEXURE-6	179
ANNEXURE-7	180
ANNEXURE-8	181
ANNEXURE-9	182
ANNEXURE-10	183

FUTURE READINGS   **184-192**

LIST OF FIGURES

Figure No.	Title	Page No.
Figure 1.1	Problems with Spread or Variation	7
Figure 1.2	Problems with Centering or Mean Value	7
Figure 1.3	Reducing Variation- Reducing Defects	9
Figure 1.4	Three Engines of Six Sigma	10
Figure 1.5	Process Design/ Re-design	11
Figure 1.6	Process Management	12
Figure 1.7	Process Improvement	14
Figure 1.8	Import Trend of Auto Parts	15
Figure 1.9	Historical Trend of Six Sigma Implementations	16
Figure 2.1	Six Sigma in Manufacturing and Service Sector	26
Figure 2.2	Six Sigma in Indian Industries	27
Figure 2.3	Top Ten Countries in Foundry Production	31
Figure 2.4	Production Trend of Various Indian Foundries	33
Figure 2.5	Bifurcation of Indian Foundries	34
Figure 2.6	Major Industrial Customers of Indian Foundries	35
Figure 2.7	Classification of Indian Foundries	35
Figure 2.8	Reasons for Avoiding Six Sigma in India	36
Figure 3.1	Research Plan	40
Figure 3.2	Regression Analysis	51
Figure 4.1	End Product (Pistons) Range	56
Figure 4.2	Market Share Descriptions	56
Figure 4.3	Proposed Framework for D-Phase	62
Figure 4.4	House of Quality	64
Figure 4.5	SIPOC Diagram for Casting Process	65
Figure 4.6	Project Charter	65
Figure 4.7	5M Diagram for Foundry Scrap	67
Figure 4.8	Proposed Framework for M-Phase	72

Figure 4.9	Process Mapping of Casting Process	74
Figure 4.10	Pictorial Presentation of Bottom Thickness Variation	74
Figure 4.11	Calculation of Cpk Value in Terms of BT Variation	75
Figure 4.12	Calculation of Sigma Value with Sigma Calculator	76
Figure 4.13	Snaps of Casting Defects	76
Figure 4.14	Pareto Chart	77
Figure 4.15	Root Cause Analysis by Cause and Effect Matrix	78
Figure 4.16	Random Data Collection Plan for Conducting Gage R&R	79
Figure 4.16	Gage R&R Measurement	79
Figure 4.17	Graphical Representation of Gage R&R Results	81
Figure 4.18	Bias Study of Immersion Pyrometer	82
Figure 4.19	Mean-Range Control Chart for Stability of Vac-Tester	83
Figure 4.20	Traditional Classification of Analysis Tools	86
Figure 4.21	Calculations for Chi-square Values and Testing	90
Figure 4.22	ANOVA for Different Coating Thickness Groups	91
Figure 4.23	Analysis of Results	92
Figure 4.24	Test for In Gate Design	94
Figure 4.25	Representations of t-Test Results	94
Figure 4.26	Normality and Fitted Line Plots of Process Variables	96
Figure 4.27	Regression Analysis	97
Figure 4.28	Residue Plots for Xs and Y	97
Figure 4.29	Interaction Plot among Three Parameters	99
Figure 4.30	Fish Bone Analysis of Defective Holes in Castings	100
Figure 4.31	Why-Why Analysis	101
Figure 4.32	Proposed Framework for Improve Phase	104
Figure 4.33	Algorithm for an Integrated Methodology of DoE	105
Figure 4.34	DoE Analysis for an Orthogonal Array	110
Figure 4.35	ANOVA Applied over DoE Model	111
Figure 4.36	Normal Plot of Effects	111
Figure 4.37	Analysis of Reduced Model	112

Figure 4.38	ANOVA Applied on Reduced Model	112
Figure 4.39	Relative Effects of Factors	114
Figure 4.40	Four Pack Chart of Fitted DoE	114
Figure 4.41	Main Effect Plot by Taking One Factor at a Time (OFAT)	115
Figure 4.42	Two Way Interaction Plots	116
Figure 4.43	Surface and Contour Plot for Two Way Interaction (BC)	117
Figure 4.44	Cube Plot for Three Way Interaction (ABD)	118
Figure 4.45	Optimization of Process Parameters	118
Figure 4.46	Results from Response Optimizer	119
Figure 4.47	Improvement in Machine Design	120
Figure 4.48	Modification in Die Design	121
Figure 4.49	Continuous Improvement in Pin Cleaning Operation	122
Figure 4.50	Training Matrix	123
Figure 4.51	Scrap Trend (From 1st July to 31st December 2010)	124
Figure 4.52	Proposed Framework for Control Phase	127
Figure 4.53	Improved Sigma Value	128
Figure 4.54	Process Capability of Improved Process	129
Figure 4.55	Casting with Increased Volume of R&R	130
Figure 4.56	Control Chart for BT Defect	131
Figure 4.57	p-Chart for Monitoring Casting scrap	131
Figure 4.58	Process Audit Sheet	133
Figure 4.59	Process Indicator Board	133

LIST OF TABLES

Table No.	Title	Page No.
Table 1.1	Research Claims on Six Sigma's Origin	5
Table 1.2	Comparison of Indian and Overseas Foundries	15
Table 2.1	Six Sigma- The Concept	19
Table 2.2	Six Sigma Themes	21
Table 2.3	Tools and Techniques Deployed in Six Sigma Programme	23
Table 2.4	Application of Six Sigma in Manufacturing Sector	24
Table 2.5	Awareness on Six Sigma Tools and Techniques in India	28
Table 2.6	Perceptions of Indian Companies on Six Sigma Benefits	29
Table 2.7	Challenges and Advantages for Foundry SMEs	30
Table 2.8	Details of Major Foundry Clusters	34
Table 3.1	Tools Used in DMAIC Phases	44
Table 4.1	Production Capability of Foundry	55
Table 4.2	COPQ Matrix	68
Table 4.3	Schedule Chart	68
Table 4.4	Random Data Collection Plan for Conducting Gage R&R	79
Table 4.5	Data Collections and Measurements for Stability Test	83
Table 4.6	Proposed Roadmap for Selection of Analytical Tools	87
Table 4.7	Plan for Analyse Phase	89
Table 4.8	Data of Chi-Square Test	89
Table 4.9	Data for One Way ANOVA	91
Table 4.10	Data for In Gate Design	93
Table 4.11	Input Data for Multi-Regression	95
Table 4.12	Interaction Plot Data	98
Table 4.13	Improvement Plan	107
Table 4.14	Description of Critical Factors with Respective Levels	108
Table 4.15	Full Factorial Design of Experiment	108
Table 4.16	Execution of Designed Experiments	109
Table 4.17	Data of BT Variation	130

NOMENCLATURE

Abbreviation / Symbol	Description
ACMA	Automotive Components Manufacturer's Association
AMP	Automotive Mission Plan
ANOVA	Analysis of Variance
BT Variation	Bottom Thickness Variation
CI	Confidence Interval
COPQ	Cost of Poor Quality
Cp	Process Capability Potential
Cpk	Process Capability Index
CTP	Critical to Process
CTQ	Critical to Quality
CWQC	Company Wide Quality Control
DF	Degree of Freedom
DFSS	Design for Six Sigma
DMADV	Define-Measure-Analyse-Design-Verify
DMAIC	Define-Measure-Analyse-Improve-Control
DoE	Design of Experiments
DPMO	Defects Per Million Opportunity
DPU	Defects Per Unit
DRO	Digital Read Out
FMEA	Failure Mode and Effect Analysis
GDP	Gross Domestic Product
Ha	Alternate Hypothesis
Ho	Null Hypothesis
KWH/T	Energy Units per Ton
Lakhs	Used in Indian Number System, 10 Lakhs= 1 Million
LSL	Lower Specification Limit
MBNQA	Malcolm Baldrige National Quality Award

MCR	Maximum Continuous Rating
MSA	Measurement System Analysis
MTPA	Million Tons per Annum
N	Data Population or Lot size
P	Probability Value
Ppk	Process Performance Capability
PPM	Parts per Million
QA	Quality Assurance
QFD	Quality Function Deployment
R&R	Repeatability and Reproducibility
R&R	Repeatability and Reproducibility
Rs	Indian Rupees (1US$= Rs 48 approx. as on 20.12.11)
RSM	Response Surface Methodology
SIAM	Society of Indian Automobile Manufacturers
SIPOC	Supplier-Input-Output-Customer
SME	Small and Medium Enterprises
SPC	Statistical Process Control
SUV	Supports Utility Vehicle
T	Target Value
TQC	Total Quality Control
TQM	Total Quality Control
USL	Upper Specification Limit
VSM	Value Stream Mapping
YB, GB and BB	Yellow Belt, Green Belt and Black Belt
Z-bench	Process Benchmark
ZDQC	Zero Defect Quality Control
Σ	Standard Deviation
$	Dollars
μ	Process (Universe) Mean

CHAPTER – 1   INTRODUCTION

1.1 Pretext

In the present market, competitors are looking for flexibility and shorter production lead times because only such a configuration of production system can fulfill the ever-changing demands of customers (De Feo, 2000). But for this it is necessary to have less scrap as it is well known that this will result into economic production (Wright and Basu, 2008). To handle these challenges, companies are being forced to adopt such strategies/techniques which can make production less costly and of optimum quality. Goh et al. (2004) highlights that by scrap reduction, one can have maximum utilization of machines or equipments, which will obviously enhance the production rate and make the overall production more feasible. So for higher productivity, 'defect reduction' will be one of the most promising and viable strategy and it will also help to cope up with the future challenges (Antony, 2002). To achieve this efficacy, Six Sigma's DMAIC approach can be efficiently used in the given production environments, particularly in foundry SMEs of India.

Over the past few decades, Six Sigma has been espoused successfully by many world-class companies. The main benefit of a Six Sigma program is the elimination of subjectivity in decision-making by creating a system where everyone in the organization collects, analyzes and displays data in a consistent way (Maleyeff and Kaminsky, 2002). Six Sigma is regarded as a well-structured methodology for improving the quality of processes and products. It helps to achieve the company's strategic goal through the effective use of project-driven approach (Khan and Al-Darrab, 2010). Six Sigma projects must be linked with business strategy and should meet the requirements of the customer (Singh & Khanduja, 2010a). Nowadays, Six Sigma has been widely adopted in a variety of industries in the world and it has become one of the most important subjects of debate in quality management. Six Sigma can help a company to achieve expected goals through continuous project improvements.

Six Sigma is an established approach that seeks to identify and eliminate defects, mistakes or failures in business processes or systems by focusing on those process performance characteristics that are of critical importance to customers (Dulluri and Raghavan, 2009). After more than two decades of successful implementation of Six Sigma methodologies at major corporations, the benefits of Six Sigma are well documented now. Although this strategy has been implemented with success in many large corporations, there is still less documented evidence of its implementation in smaller organizations, specifically in foundries. Moreover, from literature survey it appears that foundries in India are still not convinced that Six Sigma can be effectively implemented with their existing shop floor constraints. This book attempts to address the issue: Can Six Sigma be effectively implemented in Indian foundries? One of the common myths of Six Sigma that has emerged over the last few years is that it is just applicable to large corporations with immense resources and budget (Edgeman and Dugan, 2008). The purpose of this work is to test this myth and to show that a Six Sigma based business strategy is applicable to all kinds of foundries in developing countries (like India) irrespective of their size and type. The present study has raised this serious issue and tried to formulate useful frameworks for each phase of DMAIC methodology as far as Six Sigma implementation in Indian foundries is concerned.

1.2 Quality Control in Foundries through Six Sigma

The primary objective of quality control in a foundry is to improve the profitability. Another important aspect of quality control is judging the conformity of a process to established standards and taking suitable action when there are any deviations. It critically aims to improve the quality and productivity by process control, experimentation and usage of customer's feed-back (Freiesleben, 2007). This assists in developing quality consciousness in the foundry environment. Some important quality rules for improving reliability of foundry units are;

• Start with a good quality melt. Avoid turbulence of the molten metal.
• Avoid bubble entrainment by properly designed offset step pouring basin and gating system
• Avoid core blows and shrinkage by adequate venting. Avoid convection.
• Reduce segregation, particularly the channel segregation.
• Provide location points for pickup for dimensional checking and machining.

The concepts like total quality control (TQC), companywide quality and creativity (CWQC) and zero defect quality control (ZDQC) have been developed for inspection oriented approaches for identifying the problems within the end product after they occur and solving them as preventive actions. But in today's competitive scenario, SMEs cannot even afford one time scrap and need a highly capable production process with strict control on critical parameters to ensure a conforming product every time (Hamza, 2008). Since last two decades, Indian foundries are using quality techniques like TQM, QA, QFD etc. but these have been unable to generate zero defect production due to some of their inherent non compatibilities. Now in 21st century, time has come for India to change the existing quality philosophies and move towards more effective and self adaptable approaches like Six Sigma.

The elimination of these casting defects requires the collection and analysis of data (Singh & Khanduja, 2010b). There are many statistical techniques for controlling process variables, correlating the effects of variables, analyzing the problems and establishing the priorities for problem solving. It is true that the size and type of a foundry (i.e., large high-production foundry, medium or jobbing foundry) as well as the cultural thinking of its management influence the methodology to diagnose casting defects (Huq, 2006). To correctly diagnose casting defects, it is imperative to fully document the defects by illustration, description, analysis and accurate data. Attempting a corrective action without knowing exactly what the problem is, may prove very expensive (Sandhya and Costas, 2010). Once a corrective action is found, it must be implemented. In jobbing foundries, the corrective action usually takes place during the next production run but on the other hand, in the large high-production foundries the corrective action takes place almost immediately. So it can be concluded that to reduce defects or scrap in any type of foundry, one must have highly capable process (containing execution of sequence of controlled parameters) to reduce internal and external defects in the end-product. Perhaps this can be effectively achieved through strategic implementation of Six Sigma without ignoring given manufacturing constraints.

1.3 Six Sigma: Origin and Development

Origin: The origin of Six Sigma has been the subject of countless articles in journals, a series of Harvard Business School cases and many books (Raisinghani et al., 2005). Six Sigma has been very successful and is perhaps the most successful business improvement strategy in the last 50 years (Prabhushankar et al., 2008). Various authors claim that the Six Sigma as a quality initiative was started in different times like the mid-1960s, the later part of the 1970s, in 1979, in the 1980s, in the earlier part of the 1980s, in 1986, in 1987 and in the early 1990s (Kaushik, 2010). Historically, the roots of 'Sigma' as a measurement standard go back to Carl Fredrick Gauss (1777-1855), who introduced the concept of normal curve. Walter

Shewart introduced 'three sigma' as a measurement of output variation in 1922 and stated that process intervention was needed when the output went beyond this limit. The 'three sigma' concept is related to a process yield of 99.973% and represents a defect rate of 2,600 per million which was adequate for most manufacturing organizations until the early 1980s (Mahanti, 2005). Henderson and Evans (2000) state that Motorola first embarked on its Six Sigma quality initiative in the mid-1960s and the concept of implementing Six Sigma was pioneered at Motorola in the 1980s. In addition, Dedhia (2005) and Dass et al. (2006) have claimed that Bill Smith at Motorola, during the late 1970s developed the Six Sigma approach with an objective to control defects at parts per million levels instead of percentage. Schroeder et al. (2008) state that Six Sigma had its birth at Motorola in 1979, when executive Art Sundry stood up at a management meeting and proclaimed, "The real problem at Motorola is that our quality stinks". Sundry's proclamation sparked a new era with in Motorola and led to the discovery of the crucial correlation between higher quality and lower development costs in manufacturing products of all kinds.

Few authors are of the opinion that Six Sigma was started in the 1980s without being specific about the year of inception (Senapati, 2004; Antony and Banuelas, 2002; Kwak and Anbari, 2004; Man, 2002; Goh et al., 2003; Wang et al., 2004; De Koning and Mast, 2006; Chang and Wang, 2008). Chatterjee (2003) and Thirunavukkarasu (2008) are also of the opinion that Six Sigma was started in the early 1980s. According to Antony (2006), a senior engineer and scientist at Motorola's communication division introduced the concept of Six Sigma in 1986. Antony et al. (2007) claim that the first generation of Six Sigma lasted for a period of 8 years (1987-1994) and the focus was on reduction of defects. Motorola was a great example of a successful first generation company. The second generation of Six Sigma spanned from 1994 to 2000 and the focus was on cost reduction. General Electric, Du Pont and Honeywell are good examples of successful second generation companies. A majority of the authors have mentioned that Six Sigma emerged as a distinct approach to TQM in 1987 at Motorola (Klefsjö et al., 2001; Caulcutt, 2001; Wiklund and Wiklund, 2002; Dasgupta, 2003; Pandey, 2007; Black and Revere, 2006;). The summary of the papers that describe the place and year of origin of Six Sigma is presented in table 1.1. Motwani et al. (2004) have stated that the Six Sigma approach was first introduced and developed at Motorola in the early 1990s. According to Dahlgaard and Dahlgaard-Park (2006), Six Sigma methodology was first introduced in USA in 1985 at Florida Power and Light (FPL), when the company decided to apply for the Deming Prize whereas, Nonaka and von Krogh (2009) and Manual (2006) have stated that the concept of Six sigma originated in Motorola in the USA around 1985. This shows that there is no consensus on even the place of origin of Six Sigma. It can be seen that a large number of the researchers and practitioners believe and affirm that Six Sigma originated at Motorola in 1980s and was officially launched in 1987. With the exception of Dahlgaard and Dahlgaard-Park (2006), all other authors have pointed out that Six Sigma originated at Motorola.

Table-1.1 Research Claims on Six Sigma's Origin

Year	Place	Authors
1960	Motorola	Henderson and Evan, 2000; Goh et al., 2003
1970	Motorola	Harry, 1998; Man, 2002; Dedhia, 2005
1975	Motorola	Park and Kim, 2000; Moorman, 2005
1980	Motorola	Montgomery et al., 2004
1982	Motorola	Wyper and Harrison, 2000; Hammer, 2002
1983	Motorola	Antony and Banuelas, 2002; Bonilla et al., 2008
1985	FPL	Dahlgaard and Dahlgaard-Park, 2006
1986	Motorola	Kumar, 2007; Antony, 2006
1987	Motorola	Klefsjö et al., 2001; Caulcutt, 2001
1988	Motorola	Motwani et al., 2006; Raisinghani et al., 2005
1989	FPL	Wiklund and Wiklund, 2002; Dasgupta, 2003
1990	FPL	Black and Revere, 2006; Schroeder et al., 2008
1992	Motorola	Hagemeyer et al., 2006; Prabhushankar, 2009

Historical Development: The Six Sigma crusade, after Motorola has taken over scores of other companies, which are continually striving for excellence. However, this unique philosophy became well known only after GE's Jack Welch made it a central focus of his business strategy in 1995. Today, Six Sigma is the fastest growing business management system in industry. The evolution began in the late 1970s, when a Japanese firm took over Motorola in United States and promptly made drastic changes in operations management of the company. Under Japanese management, the company was soon producing TV sets with 1/20th of the number of defects they had earlier produced. Finally, Motorola recognized that its quality was awful and since then, it decided to take quality seriously. In the mid-1980s,

Motorola under the leadership of Robert W. Galvin, was the initial developer of Six Sigma (Hahn et al., 1999). He proposed 'Six Steps to Six Sigma' for process improvement and 'Mechanical Design Tolerance' for the reduction of defects to 3.4 defects per million opportunities (DPMO). Galvin was impressed by the name 'Six Sigma' because it sounded like a new Japanese car and he needed something new to attract attention. Galvin committed himself to the concept and officially launched 'Six Sigma' on 1 January 1987, which enabled Motorola to win the most coveted Malcolm Baldrige National Quality Award (MBNQA) in 1988. Then on Six Sigma became a federally registered trademark of Motorola (Prabhushankar et al., 2008; Caulcutt, 2001). During the initial period of implementation, the concentration was mainly on the statistical aspects of Six Sigma.

It was Mikel Harry and Richard Schroeder, executives at Motorola, who were responsible for the creation of the unique combination of change management and data-driven methodologies that transformed Six Sigma from a simple quality measurement and process improvement tool to what it is perceived today: a breakthrough business excellence philosophy. They developed strategies and deployment guidelines that will work in a variety of industries. They elevated Six Sigma from shop floor to boardroom. For his contribution to Six Sigma, Harry has been credited as the father of Six Sigma (Dedhia, 2005). Until 1994, Six Sigma remained a closely guarded secret at Motorola which the outside world knew about, but not how to use it. In 1995, however, CEO Gary L. Tooker decided to throw open the source code. One of the earliest to pick it up was Allied Signal, where CEO Lawrence Bossidy led the conversion (Hahn, 2005). GE began its Six Sigma programme in 1995 and has achieved remarkable results since then. The speech delivered by CEO John F. Welch, Jr. on 24 April 1996 at the GE annual meeting, declared Six Sigma as the company's quality culture, is regarded as a milestone in Six Sigma history. An exponentially growing number of global firms have launched Six Sigma programmes after GE announced its results in 1996 (Sadagopan, et al., 2005; Prabhushankar et al., 2008).

1.4 Six Sigma: A Statistical Concept

Six Sigma is a statistically based quality tool as it deals with the statistical problems. The nature of statistical problem is described in figure 1.1 and figure 1.2. Figure 1.1 shows the problem with spread or variation. Most of the quality and management problems are due to the existence of product variation. All defects and non-conformities would have vanished if variation could have been vanished. This type of problem is mainly associated with manufacturing units like foundries. The area beyond the upper specification limit and lower specification limit is corresponding to non-conforming product or processes. This area is quite high in current situation and it is desired to minimize it.

Figure-1.1 Problems with Spread or Variation

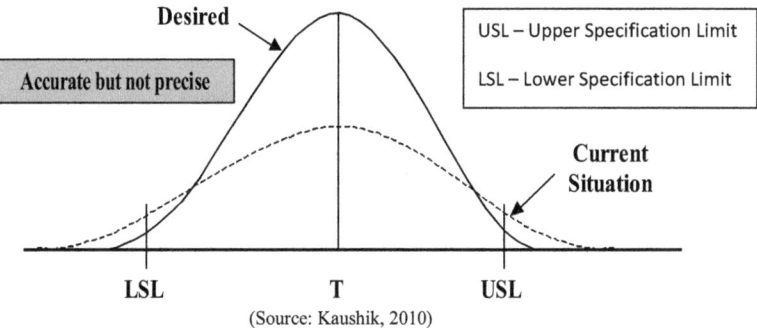

(Source: Kaushik, 2010)

Figure 1.2 shows the problem with centering or the mean value. This type of problem is mainly associated with the process industry where consumption rate is very high and to reduce the consumption rate, statisticians have to develop methodology and tools for estimating, comparing, controlling and reducing mean value.

Figure-1.2 Problems with Centering or Mean Value

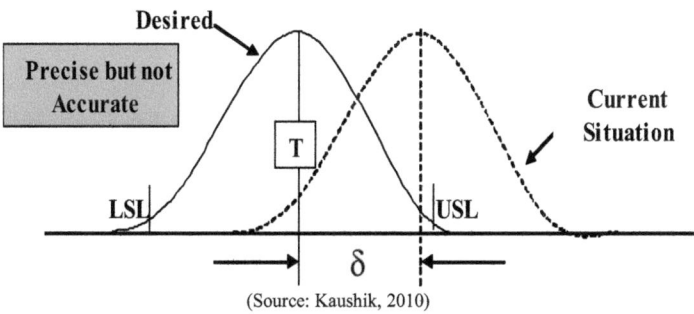

(Source: Kaushik, 2010)

Goals of Six Sigma: In statistics, sigma denotes the standard deviation of a set of data and the sigma value of the process describes the quality level of that process. A process is centered when X=T, where X is the process average or mean and T is the target value, which is typically the midpoint between the customer's upper specification limit (USL) and the lower specification limit (LSL). A process is off-centered when the process average X is not equal to target value T. The off-centering of a process is measured in standard deviations or sigma which provides a measure of variability, indicates how all data points in a statistical distribution vary from the mean (average) value.

The quality level indicates how well that process is performing (Mahanti and Antony, 2005). The higher the quality level (2σ, 3σ, 4σ etc.), the better will be the process. Sigma measures the capability of the process to perform defects free work. Six Sigma quality approach relies upon the normal distribution. Most of the output of the process will meet the specifications (call it X). But some will deviate, to varying extent, measured by the standard deviation (σ). So, some output units will be produced in the range of $X\pm 1\sigma$, some $X\pm 2\sigma$ and some in $X\pm 3\sigma$. This problem of varying quality of output units is tackled by the Six Sigma methodology in two ways. First, it widens the design width, stretching the upper and lower specification limits so that, even if the output unit is produced in $X\pm 3\sigma$ range, it will function properly. Secondly, Six Sigma provides tools to analyze and reengineer the process so that the value of sigma drops, thus if the upper and lower specification limits originally stood at $\pm 3\sigma$, they will automatically stand at $\pm 6\sigma$ now. In other words, variability of the process is to be reduced to such an extent that the value of sigma of the processes reduces to a new low value, which can be fitted between $\pm 6\sigma$ times even with the same specification limits.

The statistical representation of Six Sigma describes quantitatively how a process is performing. Six Sigma's goal is the near elimination of defects from any process, product or service-far beyond where virtually all companies are currently operating (Mahanti, 2005). Six Sigma focuses all functions on 'processes'. Every process/procedure has an expected outcome/measurement called a 'mean '. Every outcome/measurement has some variation and the measure of that variation is called sigma. Thus, the focus of Six Sigma methodology in manufacturing/process is to reduce variation as shown in figure 1.3, which also explains clearly the differences between 3σ and 6σ process.

The numerical goal of Six Sigma is reducing defects less than 3.4 parts per million (PPM), also known as defects per million opportunities (DPMO), reducing cycle time and reducing costs dramatically which impact the bottom line (Haikonen et al., 2004). Reducing variation and mean is the essence of Six Sigma and a Six Sigma defect is defined as anything outside the customer specification. Six Sigma is a disciplined, data driven approach and methodology for eliminating defects (driving towards six standard deviation between the mean and the nearest specification limit) in any process - from manufacturing to process industry and from product to service (Desai, 2006).

Figure-1.3 Reducing Variation - Reducing Defects

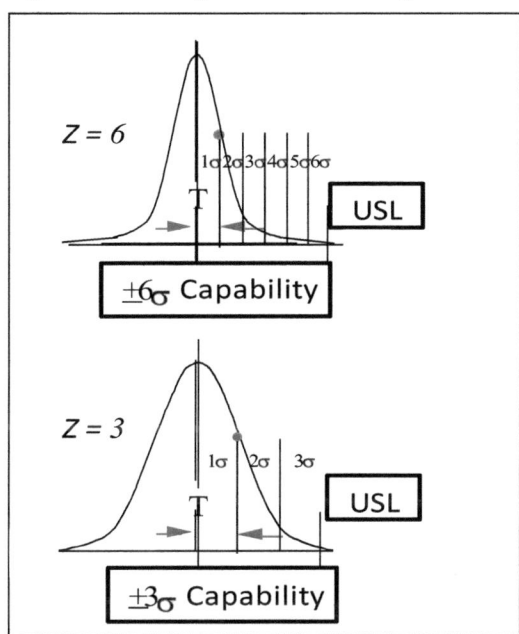

1.5 Implementation of Six Sigma

The fundamental objective of the Six Sigma methodology is the implementation of a measurement-based strategy that focuses on process and variation reduction, through application of Six Sigma projects (Hahn, 2005). In one implementation model, all Six Sigma projects run through the independent organization, making it easy to measure the impact of the changes (Feld and Stone, 2002). However, this arrangement can create a "we versus them" mentality that can undermine the effectiveness of the Six Sigma initiatives. To avoid this, other model involves a more integrated approach. In this model, Six Sigma is incorporated into every employee's job and hence makes it more challenging to measure the impact of Six Sigma (Goh et al. 2006). It helps to create a culture in which a commitment to quality and excellence is pervasive. Basically Six Sigma implementation fuels three engines- Process Design/Redesign, Process Management and Process Improvement (Kaushik and Khanduja, 2008), as shown in figure 1.4.

Figure-1.4 Three Engines of Six Sigma

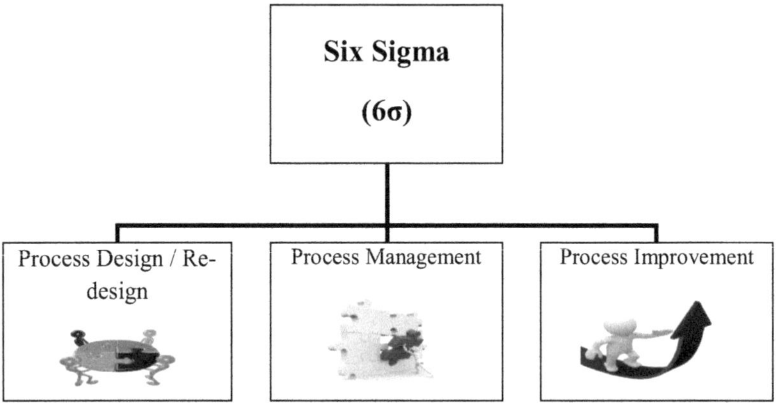

Process Design/Redesign: It ensures basic need in the business where it is required to replace rather than repair, one or more core processes. If no process exists or if the existing processes are deemed beyond repair, then 'Design for Six Sigma' (DFSS) or 'Six Sigma Design' (SSD) methods are used to create effective processes. In process design, Six Sigma principles are used to create revolutionary new processes, goods and services built around customer requirement and validated by data and tests. The five phase DMADV (Define, Measure, Analyze, Design, and Verify) approach (as explained in figure 1.5) is used for the Process Design. DMADV can be explained as:

- ***Define:*** This involves defining customer requirements and goals for the process/product/service. This phase identifies the project goals and specifies the scope of customer requirements.

- *Measure:* This phase identifies the CTQs for the design (Product or Process). It gathers valid baseline information and establishes design goals. This phase is also necessary to identify and rank the design opportunities.
- *Analyse:* This phase begins with high level concept design and establishment of system architecture elements or building blocks. It assesses process design and refines future requirements.

Figure-1.5 Process Design/Re-design

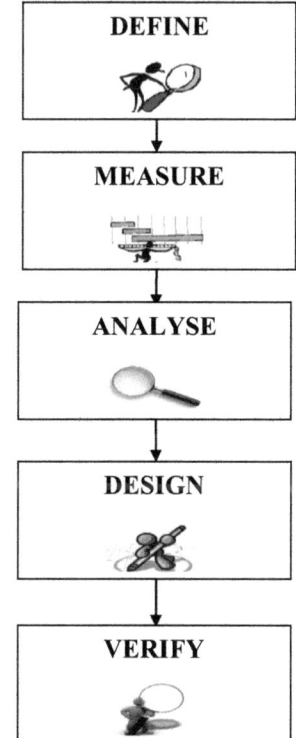

- *Design:* The design solution for delivering CTQs with associated quality level is determined through prediction of quality during this phase. This phase not only designs and implements new processes/product /services but it also designs workflow principles, apply creativity and challenge assumptions.

- *Verify:* This phase ensures that design is verified. Learnings are fed back into predictive models, methods, data and assumptions to improve method for next design. This phase verifies results and maintains performance.

Process Management: 'Process Management' means that a focus on managing processes across the organization replaces managing individual functions by different internal departments. It is the most evolutionary of the three because it involves changes in culture and management throughout the organization that must accompany Six Sigma efforts if their full power is to be realized. It is a fundamental makeover for making an organization structured and managed. In general, process management includes DMAC (Define, Measure, Analyze, Control) strategy, as shown in figure 1.6. DMAC can be explained as;

Figure-1.6 Process Management

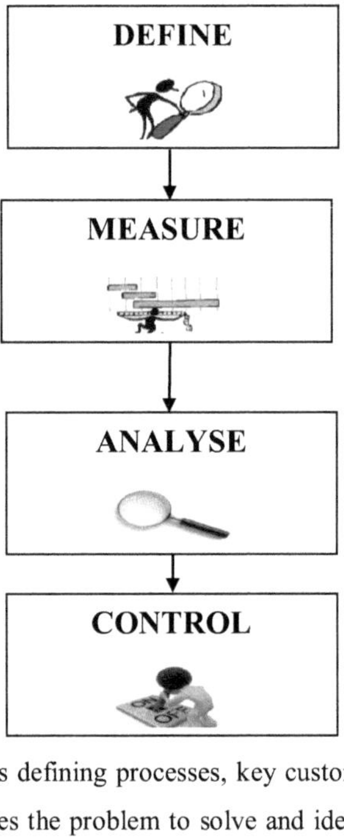

```
DEFINE
  ↓
MEASURE
  ↓
ANALYSE
  ↓
CONTROL
```

- *Define:* This phase involves defining processes, key customer requirements and process "owners." This phase defines the problem to solve and identify the impact and potential benefits keeping in mind the customer requirements.

- *Measure:* This phase requires identifying the critical-to-quality (CTQ's) characteristics of the process in order to focus Six Sigma on areas that will have greatest impact on customer satisfaction.
- *Analyse:* This phase involves analyzing data to enhance measures and refine the process management mechanisms.
- *Control:* It involves controlling performance through ongoing monitoring of inputs/operations/outputs and responding quickly to problems and process variations. Process management is work that business leaders do to improve processes for managing business.

Process Improvement: It refers to a strategy of finding solutions to eliminate the root cause of performance problem in the existing processes (Bendell, 2006). Its role is to help the management to produce maximum by using minimum input which create improvement programs. Process improvement also finds the critical Xs (causes) that create the unwanted Ys (defects) produced by the process. If there is an existing process that is not meeting customer specifications, then using Six Sigma five phase methodology DMAIC (Define, Measure, Analyze, Improve, Control) as shown in figure 1.7, that process can be improved and made more effective and efficient. DMAIC can be explained as:

- *Define:* Define the problem and what the customer requires. The define phase sets the expectation of the improvement of project and maintenance of focus of Six Sigma strategy on customer's requirements. There are many tools used in Six Sigma methodology for defining the problem and some of these are; QFD, FMEA, Logic Tables and Pareto Charts.
- *Measure:* The measure phase identifies the defects in the product, gathers valid baseline information about the process and establishes improvement goals.
- *Analyse:* This phase examines the data collected in order to generate a prioritized list of sources of variation. It is the key component of any defect reducing program. This is the stage at which new goals are set and route maps created for closing the gap between current and target performance level. Statistical tools as well as conventional quality techniques like Brainstorming, Root-Cause Analysis, Normality Analysis, Process Capability, Fish Bone Diagram, Pareto Analysis etc. may be used for carrying out the analysis.

- ***Improve:*** The optimal solution for reducing variation or mean is determined and confirmed in the improve phase. The objective of this phase is to confirm the key process variables by use of brainstorming and action workouts. This phase helps to identify and quantity the key process variables and their influence on CTQs. It determines acceptable limits to reduce the number of defects in the process. This step may involve the use of a variety of statistical methods and tools to determine high priority attrition variables and to develop and/or redesign functions that impart product performance and success.

Figure-1.7 Process Improvement

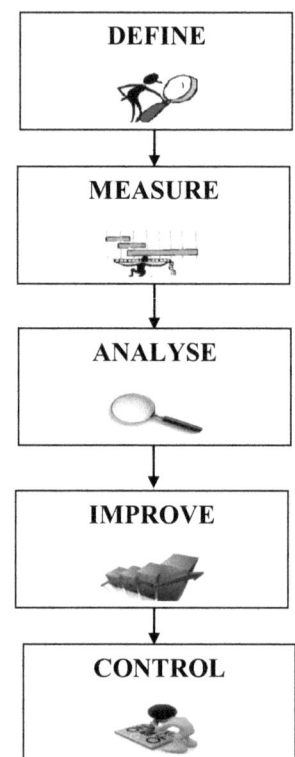

- ***Control:*** The final stage of Six Sigma implementation is to hold the gains that have been obtained from the improve stage. Hence in this stage the new process considerations are documented and frozen into systems so that the gains are permanent. This phase emphasizes process capability and implementing various process controls to make sure that the modified process stays within acceptable limits.

1.5 Motivation for the Present Book

The performance of Indian foundries has been highly dismal as shown in table 1.2, which compares their performance indices with those of other countries. Poor operation management, lack of serious R&D support, low capacity utilization etc. have been the major reasons to this low productivity (Zhan, 2008; Wyper and Harrison, 2000; Yeh, 2007).

Table- 1.2 Comparisons of Indian and Overseas Foundries

Performance Norms	Indian Foundries	Overseas Foundries
Production per annum (%)	8 to 9	92 to 93
Rejection (%)	10 to 25	2 to 12
Capacity Utilization (%)	45 to 55	60 to 75
Productivity (T/Man/Year)	12 to 20	100 to 120 (Japan) 50 to 55 (Germany) 10 to 30 (China)
Energy Requirement (KWH/T)	700 to 900	400 to 600

To further aggravate this scenario, Automotive Mission (AMP) Plan 2006-2016 of India has put tremendous pressure on foundry units because foundry units of the country are finding it difficult to meet the steep demand of Indian auto industry (refer figure 1.8). Due to this, import rate of castings is increasing and future trend seems to be alarming. Hence there is dire need of devising serious strategic measures to increase the productivity levels of Indian foundries.

Figure-1.8 Import Trend of Auto Parts

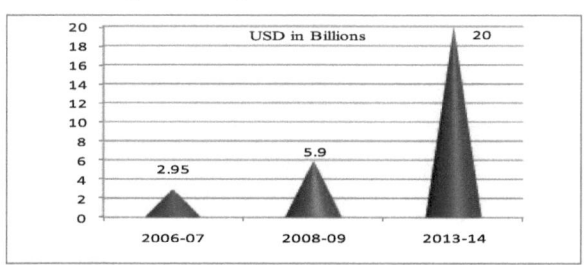

(Source: Panchal 2010, Indian Foundry Journal)

The present book is an exploration in this direction as this work dwells on testing some Six Sigma imperatives for enhancing production efficiency in Indian foundry environments.

Literature review also shows that Six Sigma research has been mostly empirical in nature which reinforces the use of real-world data. Case study was the dominant approach in Six Sigma research and this is perhaps due to the fact that quality problems in manufacturing and service contexts are usually treated as a case in terms of documentation and analysis. Figure 1.9 defines the growing gap over the years between case study method and other research methods, particularly survey research. Case study method is used to document and analyze Six Sigma implementation in particular contexts; industry, service, process or phase of a specific project. In addition, the lack of implementing Six Sigma tools and methodologies across a wide range of processes or organizations makes the use of survey approach impractical. The graph in figure 1.9 shows that case study based approach has been well acceptable and successful since 1992, as far as Six Sigma concept and its implementation are concerned. From 2004 to 2008, researchers seem to be using this approach exponentially as compared to survey based and review based frameworks by analyzing the benefits and authenticity of case study based works in the field of Six Sigma.

Figure-1.9 Historical Trend of Six Sigma Implementations

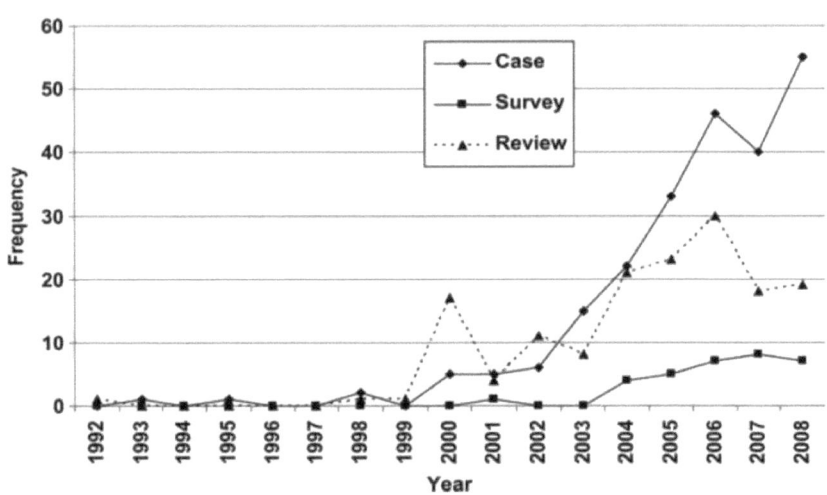

(Source; Aboelmaged, 2009)

The present book has taken this case study based approach to achieve the pre defined goals and objectives relating to Six Sigma implementation in Indian foundries.

1.6 Objectives of the Book

Primarily the major objectives of the present study are:

- To formulate a comprehensive strategy for the successful implementation of Six Sigma in Indian foundry industry without ignoring its existing constraints.
- To provide standard step by step framework to execute different phases of DMAIC project in Indian foundry environments.
- To devise a strategy for selection of right tool/technique for right problem.
- To demystify the different myths prevailing about Six Sigma for Indian SMEs, specifically for foundry units.

Exercise

1. What do you mean by Six Sigma?
2. What is the difference between Six Sigma and TQM?
3. Explain the concept of Standard Deviation?
4. Differentiate „Defect" from „Defective"?
5. What do you mean by Scrap? How it is different from non-conforming?
6. Define Cpk? What is the difference between Cpk and Ppk?
7. Discuss three engines of Six Sigma?
8. Demonstrate various methodologies or approaches of Six Sigma?
9. What do you mean by USL and LSL?
10. Write a short note on AMP Plan of Government of India?
11. Compare the Indian foundry metrics with overseas foundries?
12. Describe the Origin of Six Sigma in detail?
13. Name at least four industries which are using Six Sigma in India?
14. Define the significance of ongoing certificate programmes in market?
15. Discuss the term Quality Control?

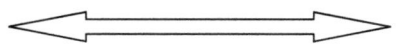

CHAPTER – 2   ANTECEDENTS OF SIX SIGMA

The following issues were predominantly addressed during an extensive literature review on various aspects of Six Sigma and its application to foundry industry. These issues primarily covered in various journals, reference manuals, handbooks, text books, e-resources etc are:

- Evolution of Six Sigma
- Six Sigma in Manufacturing Sector
- Six Sigma in Indian Industries
- Six Sigma in Foundry Sector

2.1 Evolution of Six Sigma

Six Sigma was first espoused by Motorola in 1987 and was taken up by Allied Signal in 1991. In 1995, Jack Welch, the CEO of General Electric successfully implemented Six Sigma in many processes and documented significant gains in process and financial results (Coronado and Antony, 2002; Pfeifer et al., 2004). The simplest definition for Six Sigma is to eliminate waste and to make mistake proof the processes that create value for customer. The elimination of waste led to yield improvement and production quality; higher yield increased customer satisfaction. Ehie and Sheu (2005) have indicated that the concept of Six Sigma is the development of a uniform way to measure and monitor performance and set extremely high expectations and improvement goals. Aggogeri and Gentili (2008) have concluded that Six Sigma is a highly disciplined process that helps an organization to focus on developing and delivering near-perfect products and services. The Six Sigma methodology of measuring and monitoring performance issue deals with a variety of statistical applications. The objective of Six Sigma is to enhance the sigma level of performance measures that reflect the needs of the customer. In addition, the Six Sigma level of performance means a product defect rate of 3.4 per million opportunities for error.

2.1.1 Definition of Six Sigma

With more than two decades of successful implementation of Six Sigma methodologies at major corporations, the success and benefits possible with Six Sigma are well documented. Although Six Sigma has been implemented with success in many large corporations, there is still less documented evidence of its implementation in smaller organisations. Moreover, from literature survey it has been found that SMEs are still not convinced that Six Sigma can be effectively implemented by them. A chronological review on the concept of Six Sigma is given in table 2.1.

Table-2.1: Six Sigma- The Concept

SN	Author(s)	Six Sigma as a Concept
1.	Behara et al. (1995)	Six Sigma is a statistical measure of the performance of a process or product. It is used as a quality control mechanism, which seeks to reduce defects or variations in a process to 3.4 per million opportunities thereby optimizing output and increasing customer satisfaction.
2.	Antony and Kaye (1996)	Six Sigma as a powerful management strategy has evolved from being exclusively about the original goal of a target of less than 4 failures or defects or errors per million opportunities, to encompass a broad range of approaches for incorporating quality into products and services from the early design and development stages and throughout their lifetimes
3.	Husband (1997)	Six Sigma is a well-established approach that seeks to identify and eliminate defects, mistakes or failures in business processes or systems by focusing on those process performance characteristics that are of critically important to customers.
4.	Hendricks and Kelbaugh (1998)	Six Sigma is a business improvement strategy that seeks to find and eliminate causes of defects or mistakes in business processes by focusing on outputs that are important to customers."
5.	Hahn et al. (1999)	Sigma (σ) is the Greek letter associated with standard deviation. However, in Six Sigma it takes on various definitions and interpretations such as, a metric of comparison, a benchmark comparison, a vision, a philosophy, a methodological approach or a goal.
6.	Sanders and Hild (2000)	Six Sigma is emerging approach to quality assurance and quality management with emphasis on continuous quality improvements. The main goal of this approach is reaching level of quality and reliability that will satisfy and even exceed demands and expectations of today's demanding customer.
7.	Ingle and Roe (2001)	Six Sigma is a methodology accompanied with highly structured processes using efficient statistical approaches for acquiring, assessing and applying the customer, competitor, enterprise and market intelligence to produce superior product, process and enterprise innovations and designs with the goal of creating a sustainable competitive advantage.
8.	Antony (2002)	A powerful multifaceted approach empowered by statistical tools for ensuring defect free products through continuous process improvement. It is customer-oriented, structured, systematic, proactive and quantitative approach for continuous process improvement in the business processes of an organization to ensure improved quality and low cost.
9.	Banuelas and Antony (2002)	Six Sigma is a comprehensive program for managing a business that emphasizes an intelligent blending of the wisdom of the organization with proven statistical techniques to improve both the efficiency and effectiveness of the organization in meeting needs.
10.	Hong and Goh (2003)	Six Sigma has been considered a philosophy that employs a well-structured continuous improvement methodology to reduce process variability and drive out waste. It focuses on the root causes of business process/problems to reduce the variation of occurrences around the mean value of the process data.
11.	Markarian (2004)	The Six Sigma philosophy is a quality-focused program that requires process design that can accept twice the normal variation of ± 3 sigma in a process, even if the process mean shifts by as much as ± 1.5 sigma. Thus, the Six Sigma approach to quality ensures that a maximum of 3.4 parts per million are defective in each step of the process.

12.	Marti (2005)	Six Sigma is set of collective plans, activities, and events designed to ensure that products, processes, and services will satisfy customer needs. In brief, Six Sigma is a customer-focused approach to business that provides an overall framework for quality management.
13.	Kwak and Anbari (2006)	Six Sigma is a high-performance, data-driven method of improving quality by the removal of defects and their causes in business process. Critically, it concentrates on those outputs that are important to customers.
14.	Black and McGlashan (2006)	Six Sigma is an organized and systematic method for strategic process improvement and new product and service development that relies on statistical methods and the scientific method to make dramatic reductions in customer defined defect rates.
15.	Black and Revere (2006)	A quality movement, a methodology and a measurement. As a quality movement, Six Sigma is a major player in both manufacturing and service industries throughout the world. As a methodology, it is used to evaluate the capability of a process to perform defect-free, where a defect is defined as anything that results in customer dissatisfaction
16.	McCarty and Fisher (2007)	Six Sigma is a disciplined methodology that uses data and statistical analysis to measure and improve a company's operational performance. It focuses on identifying and eliminating "defects" in business processes and has produced hundreds of millions of dollars in new profitability in a wide variety of industries
17.	Chakrabarty and Tan (2007)	Six Sigma is a formal methodology for measuring, analysing, improving and then controlling or 'locking in' processes. This statistical approach reduces the occurrence of defects from a three-sigma level or 66 800 defects per million (the average for most companies) to a Six Sigma level or less than four defects per million.
18.	Hamza (2008)	Six Sigma is for a broader institutional malady, while kaizen is for fixing small problems in a matter of days. Six Sigma is a quality initiative that employs statistical measurements to achieve 3.4 defective parts per million – the virtual elimination of errors.
19.	Snee (2009)	Six Sigma is a business philosophy of driving behavior by making an organization's values explicit in its compensation system and a business strategy of cutting costs and boosting customer satisfaction. Six Sigma is information driven methodology for reducing waste, increasing customer satisfaction and improving processes with a focus on financially measurable results.
20.	Pepper and Spedding (2010)	Six Sigma is a relentless and rigorous pursuit of the reduction of variation in all critical processes to achieve continuous and breakthrough improvements that impact the bottom-line or top-line of the organization and increase customer satisfaction, commitment and loyalty.
21.	Daniel (2011)	Sigma is basically a Greek term for variation. Six Sigma is an extremely well structured program whose prime objective to improve the business processes by minimizing the variations or defects during the production process. It is defined as 3.4 defects per million products.

2.1.2 Major Themes Emphasized

Table 2.2 depicts certain themes whose synergy with Six Sigma has created considerable interest among the researchers during the last 20 years. The literature survey has been performed to cover all articles related to Six Sigma published in international journals from 1995 to 2010. Table 2.2 provides a comprehensive list containing the themes and some classified references for each theme.

Table-2.2 Six Sigma Themes

S.N	Themes	References	Review of Work
1.	Quality approaches	Lochner (1990); Hoerl (1998); Malhotra (1999); Hild et al. (2000); Klefsjo et al. (2001); Antony (2002); Revere and Black (2003); Ribardo and Allen (2003); Haikonen et al. (2004); Hong and Goh (2004); Ferng and Price (2005); Ehie and Sheu (2005); Klefsjo et al. (2006); Andersson et al. (2006); Black and McGlashan (2006); Al-Mishari and Suliman (2008); Aggogeri and Gentili (2008); Wu and Lin (2009); Parsad and Reddy (2010); Zu et al. (2011); Singh & Khanduja (2010c); Singh & Bakshi (2012).	Six Sigma advantages involve the focus on financial and business results, use of a structured method for process improvement or new product introduction, use of specific metrics such as DPMO, critical-to-quality (CTQ) and use of a significant number of full-time improvement specialists. Six Sigma authors asserted that Six Sigma is not an alternative to TQM. Even though most Six Sigma tools and techniques are already being applied in the TQM field and both approaches preach that continuous improvement of quality is essential to business success, there is a vital distinction between them. Hence, the impression raised by some researchers that Six Sigma could be easily implemented in a company that implements TQM is still debatable.
2.	Six Sigma Education	William and Carley (1996); Tomkins (1997); Maleyeff and Kaminsky (2002); Man (2002); Mitra (2004); Ho et al. (2006); Stevenson and Mergen (2006); Mehta et al. (2006b); Rao and Rao (2007); Weinstein et al. (2008); Gijo and Scaria (2010); Horacio Soriano-Meier et al., (2011).	These articles have suggested that all students of management must leave the institution as certified black belts and have revealed a positive feedback from students regarding a course on Six Sigma taught at university levels. It is believed that a Six Sigma framework provides an excellent platform for integrating statistical, management and technical tools and skills into the curricula of engineering and business schools to help students tackle business problems in organizations.
3.	Methodology DAMIC/ DFSS	Fowlkes (1995); Stewart (1997); Wyper and Harrison (2000); Henderson and Evans (2000); Holtz and Campbell (2004); Lipscomb and Lewis (2004); Banuelas et al. (2005); Bandyopadhyay and Lichtman (2007); Edgeman and Dugan (2008); Kaushik and Khanduja (2008); Gizo and Scaria (2010).	The second largest part of Six Sigma literature deals with the theorization and application of Six Sigma methodologies. There are two major improvement methodologies in Six Sigma. Several studies have shown successful cases of DMAIC application in a variety of contexts such as healthcare, thermal power plants, retailing, financial services and manufacturing process. In contrast, the second methodology, design for Six Sigma (DFSS), is used for new processes or when the existing processes are unable to achieve business objectives.
4.	Lean and Six Sigma	Tomkins (1997); Hoerl (1998); Yusof and Aspinwall (1999); Hahn et al. (1999); Park and Kim (2000); Sharma (2003); Brett and Queen (2005); Bendell (2006); Chang and Su (2007); Bonilla et al. (2008); Garg (2010); Andre A de Waal et al., (2011). Singh et al. (2011); Singh & Khanduja (2011c); Joshi and Singh (2013)	Recent Six Sigma studies have focused on the relationship between Six Sigma and lean production or on the implementation of the new labeled concept "Lean Six Sigma". Lean and Six Sigma complement each other and represent a powerful framework for eliminating process waste and variation when used together. Lean production is primarily concerned with eliminating waste and reducing cycle time in processes.

5.	Belt system	Harry (1998); Henderson and Evans (2000); Hoerl (2004); Ingle and Roe (2001); Rasis et al. (2002a, b); Haikonen et al. (2004); Motwani et al. (2004); Gowen and Sadagopan et al. (2005); Lee-Mortimer (2006); Savolainen and Haikonen (2007); Schroeder et al. (2008); Ho et al. (2008); Gibbons and Burgess (2010); Thomas and Barton (2011); Joshi et al. (2013).	A black belt is a full-time team leader dedicated to the Six Sigma initiative. Black belts are equipped with expertise in using the Six Sigma methodology and statistical analysis techniques for process improvement. Individuals at the highest level of expertise in Six Sigma methodologies are called master black belts. Literature on Six Sigma belt system focuses mainly on belts training and attributes Most Six Sigma organizations adopt the hierarchical level of black belt and green belt systems. They teach, coach and mentor the lower-level black belts and green belts.
6.	Benefits	Behara et al. (1995); Carnell and Lambert (2000); Hutchins (2000); Feld and Stone (2002); Johnson and Swisher (2003); Kuei and Madu (2003); Ganesh (2004); Snee (2004); Chen et al. (2005); Das et al. (2006); Freiesleben (2007); Kumar et al. (2007); Sahoo et al. (2008); Agarwal and Bajaj (2008); Snee (2010); Bhardwaj et al. (2010b); Khare (2011); Singh and Khanduja (2014).	The most cited benefit of Six Sigma in the literature is customer satisfaction and further successful application of Six Sigma quality is positively correlated with better financial performance and profit generation. When Six Sigma is implemented successfully, it will offer a disciplined approach for improving effectiveness and efficiency in a broad range of businesses. In the manufacturing context, Six Sigma benefits are related to various areas such as reduction in process variability, reduction in in-process defect levels etc.
7.	Organizational Change	Carnell and Lambert (2000); De Feo and Bar-El (2002); Motwani et al. (2004); Thawani (2004); Sadagopan et al. (2005); Craven et al. (2006); Rajamanoharan and Collier (2006); Davison and Al-Shaghana (2007); Immaneni et al. (2007); Fazzari and Levitt (2008); Lok et al. (2008); Irani (2011); Lagrosen et al. (2011); Singh & Bakshi (2014)	Many authors have seen Six Sigma as an organizational change vehicle that possesses a culture of accountability, quality, and innovation and suggested that Six Sigma should be viewed as an organization change process. The view will improve Six Sigma implementation through identifying what needs to be changed and boost change management process itself. It revealed that Six Sigma has been deployed strategically to change the culture of organization through inculcating process control discipline in business context.
8.	Challenges	Snee (1993); Hoerl (1998); Sanders and Hild (2000); Antony et al. (2002); Feld and Stone (2002); Hammer (2002); Goh and Xie (2004); McAdam and Lafferty (2004); Senapati (2004); McAdam et al. (2005); Gijo and Rao (2005); Goh et al. (2006); Mehta et al. (2006a); Kwak and Anbari (2006); Kumar (2007); Yeung, S. (2007) Kumar et al. (2008); Liu et al. (2008); Shahabuddin (2008); Prabhushankar et al. (2009); Azadeh et al. (2010); Bhardwaj et	To Bridge the gap between theory and practice of Six Sigma, research has been given more attention by Six Sigma researchers. The various challenges and limitations, which have been illustrated by corresponding papers are as follows: - The 1.5 sigma shift resulting in a 3.4 DPMO does not make sense in service processes. - The impact of leadership styles on Six Sigma success needs more research. No unified standards have been accepted regarding the contents of belt training. - The relationship between COPQ and its financial impact in SMEs needs further research since SMEs are hardly considering quality costs. - Availability of quality data is still a great challenge in

		al. (2010a); Lagrosen (2011); Singh and Sodhi (2014).	Six Sigma projects.
9.	Success factors	Sanders and Hild (2000); Antony and Banuelas (2002); Coronado and Antony (2002); Byrne (2003); Wessel and Burcher (2004); Frings and Grant (2005); Gowen (2005); Laosirihongthong et al. (2006); Yeung, S. (2007); Chakrabarty and Tan (2007); Shanmugam (2007); Yang et al. (2008). Taylor and Taylor (2009); Chaudhary and Jain (2010); Irani (2008); Bhasin (2011)	Key factors for success or failure during Six Sigma implementation have always been subject to intensive literature. The most cited success factors in Six Sigma literature include the following: - Strong top management involvement and commitment. - Selection of Six Sigma projects. Changing organizational culture. - Aligning Six Sigma projects to corporate business Objectives. Cross-functional team working. - Effective communication. Infrastructure (both organizational and IT-infrastructure).
10.	Organizational learning	Wiklund and Wiklund (2002); Jeffery (2005); Ricondo and Viles (2005); Box (2006); Savolainen and Haikonen (2007); Nikhil et al. (2007); Khare et al. (2007); Lin et al. (2008); Nikhil (2008); Gijo and Scaria (2010); Eisenhower (2011); Singla and Singh (2014).	It includes factors that are essential for improving organizational learning and for stimulating the competence, development and motivation among personnel. Few studies have looked at the link between Six Sigma and organizational learning from a perspective that Six Sigma methodologies are mature enough to be integrated with different learning approaches.

(Source: Aboelmaged, 2009)

2.1.3 Six Sigma Tools and Techniques

There are several tools and techniques available to top managers that can be used during Six Sigma projects in achieving the desired objectives. Table 2.3 summarizes some common Six Sigma tools and techniques used during implementing Six Sigma projects.

Table-2.3 Tools and Techniques Deployed in Six Sigma Programme

Author(s)	Tools/Techniques	Description
Rasis et al. (2002); McAdam and Evans (2004); Banuelas et al. (2005); Desai (2006)	Histogram, Contingency Tables, Scatter Diagrams	The great deal of Six Sigma literature has focused on Six Sigma tools and techniques. They can be described as practical methods and skills employed by Six Sigma project teams to tackle quality related problems for fostering performance improvement. While Six Sigma tool has a specific role and is often narrow in focus, Six Sigma technique has a wider application and requires
Banuelas and Antony (2003); Snee (2009); Desai (2006); Su et al. (2006); Dua et al. (2014); Jolly and Singh (2014)	Run Chart, Moving Average Charts, Attribute charts	
Henderson and Evans (2000); Banuelas and Antony (2003); Thomas and Barton (2006)	Statistical Analysis, Theory of Constraints, Gap Analysis	
Raisinghani et al. (2005); Kumi and Morrow (2006), Tang et al. (2007)	FMEA, Process Modeling, Value Stream Mapping.	
Feld and Stone (2002); Freiesleben (2006); Kuruuzum and Akyuz (2009); Singh and Joshi (2015); Singh et al. (2014)	Pareto Analysis, Process Capability Assessments,	
Yilmaz. and Chatterjee (2000); Taner et al. (2007), Dulluri and Raghavan (2009)	Cause and Effect Matrix and Quality Function Deployment,	

References	Tools/Techniques	
Rasis et al. (2002); Raisinghani et al. (2005); Sokovic et al.(2006)	Measurement System Analysis, Process Control	specific skills, creativity and training. Examples of Six Sigma tools include Pareto analysis, root cause analysis, process mapping or process flow chart, Gantt chart, affinity diagrams, run charts, histograms, quality function deployment (QFD), Kano model, brainstorming, etc. Examples of Six Sigma techniques include statistical process control (SPC), process capability analysis, suppliers-input-process-output-customer (SIPOC), SERVQUAL, benchmarking, etc. Moreover, a Six Sigma technique can utilize various tools. For example, statistical process control (SPC) is a technique that utilizes various tools such as control charts, histograms, root cause analysis, etc. Methods, Tools and Techniques are vital to the success of any Six Sigma project whether DFSS or DMAIC.
Antony (2002); Benedetto (2003); Knowles et al. (2004); Raisinghani et al. (2005); Singh and Khanduja (2010e)	Design of Experiments, Response Surface Methodology (RSM)	
Maleyeff and Kaminsky (2002); Rasis et al. (2002); Ehie and Sheu (2005)	Control Charts, Cumsum Diagrams	
Schroeder et al. (2008); Hong and Goh (2003); Huq (2006); Hutchins (2000)	Fish Bone charts, Pareto Charts, Scatter Diagram	
Henderson and Evans (2000); Hwang (2006); Ingram (2000); Juras et al. (2007)	t-Test, Chi-square Test, Scatter plot, ANOVA,	
Patterson et al. (2005); Isaacson (2008); Hsu et al. (2008); Johnston et al. (2008)	Artificial Intelligence, Fuzzy Logic, Neural Networks	
Banuelas and Antony (2003); Rajagopalan et al. (2004); Hu et al. (2005); Singh & Khanduja (2012c); Singh and Khanduja (2011e); Singh et al. (2012)	Process Mapping, Flow Charts, Multi-varies Analysis.	
Maleyeff and Kaminsky (2002); Raisinghani et al. (2005); Hu et al. (2005)	Capability Analysis, Design of Experiments, Poka-Yoke	
Hu et al. (2005); Antony (2008); Ho and Chuang (2006);Schroeder et al. (2007)	Statistical Process Control, High Level Process Mapping	
Hu et al. (2005); Sokovic et al. (2006); Craven et al. (2006); Camgoz-Akdag (2007)	Boxplot, Individual Value Plots, Benchmarking	
Sarda; (2007); Drenckpohl et al. (2007); Hamza (2008); Frank (2003); Singh & Khanduja (2010d); Bakshi et al. (2012)	QFD, TPM, Kanban, Kaizan, 5S, TQM, Regression analysis	
Rasis et al. (2002); Banuelas and Antony (2003); Mahanti (2005); Thomas and Barton (2006), Kumi and Morrow (2006)	SIPOC Analysis, Activity Network Diagram, Theory of Constraints	
Wu and Lin (2009); McAdam and Evans (2004); Sokovic et al. (2006); Singh & Khanduja (2012d); Sodhi et al. (2012)	Matrix Plot, Affinity Diagram, Prioritization Grid	

2.2 Six Sigma in Manufacturing Sector: This concept has been widely used in manufacturing sector from last 25 years as company like Motorola has been improving its processes since 1986 by using its defect reduction approach. Similarly manufacturing giants like General Electric and Honey Well have been using it as cycle time reduction tool, since 1996. Other well known companies like Ford, Caterpillar, Our lady of Lourdes medical centre, LG and Samsung etc. are also practicing Six Sigma as a quality improvement technique in their respective manufacturing processes from 1999. Table 2.4 cites major works of the researchers related to application of Six Sigma in manufacturing sector during the past decade.

Table-2.4: Application of Six Sigma in Manufacturing Sector

SN	Author(s)	Company /	Parameters	Achievements
		LARGE SCALE COMPANIES		
1.	Bañuelas and Antony (2002)	Motorola	In-process defect level. Improve productivity.	200-times scrap reduction; saved $15billions per annum.
2.	Hong and Goh (2003)	General Electric	Turnaround time at repair shops (MTBF).	62% reduction; $2billions per annum profit earned.
3.	Kwak and Anbari (2004)	Honeywell	Concept to shipment cycle time reduction.	Reduced from 18 months to 8 months; $1.2 billions saved.
4.	Montgomery et al. (2005)	Ford	Vehicle rollovers, recalls and production.	Reduced defects by 70%; profit of $1billions per annum
5.	Koning and Mast (2006)	Caterpiller	Quality improvement, Reduced cost structure.	Reduced defects by 75%; Not Reported the savings
6.	Kleasen (2007)	Our lady of Lourdes	Availability of beds and delays in emergency.	Bed availability reduced from 267 to 235 minutes.
		MEDIUM SCALE COMPANIES		
1.	Ingle and Roe (2001)	Medium sized welding unit	Optimization of Arc - welding process.	Joint strength is increased by 26% and scrap work is reduced.
2.	Does et al. (2002)	Bulb making unit of SME	Improve the process, reduced the breakage.	Sigma level increased from 3.1 to 4.5, saving 2.5bn $ / annum.
3.	Sharma (2003)	Baxer Battery Limited	Life of battery, wastage control and pollution.	Customer satisfaction and 12% annual increased in sale.
4.	Holtz and Campbell (2004)	Ford Motors private Ltd.	Maintenance and repair improvement MTBF.	Savings of $ 60000 achieved in six months. MTBF decreases.
5.	Hu et al. (2005)	A medium scale IC engine unit	Improve the Cpk. Scrap reduced by 12%.	Process capability improved from 1.1 to 2.9.
6.	Sokovic et al. (2006)	A gravity die casting unit	Casting scrap reduced from 23% to 11%.	40% reduction in production cost with annual savings.
7.	Aggogeri and Gentili (2008)	Cranberry Drinks Ltd.	Improvement in packing & time saving.	DPMO level improved from 3011 to 178 only. Reduce time.
8.	Gadallah (2009)	Wilson Tools Private Ltd.	Shorten the treatment time & energy cost.	Roughly $10000 per year savings. 2% reduction in time.
9.	Siddiqui and Yang (2010)	A copper wire manufacturing	Quality improvement in rolling operation.	Defect are decreased by 19% with in nine months of DMAIC
10.	Zu et.al, (2011)	An Electronic manufacturing	Defects in surface mount soldering &LED	DPU reduction from 108209 to 60975 and overall scrap 7.48%.
		SMALL SCALE COMPANIES		
1.	Thomas and Barton (2006)	Orange box Limited	Office seat and design of furniture, chairs etc.	£60000 saving per annum. Reduction in material wastage.
2.	Rowe (2006)	washing machine	Failure of drum bearings, RPM and load	A huge reduction in customer complaints and dis-satisfaction.
3.	Desai (2006)	A Steel valves and fittings	Production planning and improvement.	50% improvement in delivery failures at a level of 2 sigma.
4.	Sarda (2007)	Two wheeler automobile	Gear pinion noise, short life period of gears.	Noise reduced by 13%, Sigma level increased to 3.944.
5.	Juras et al. (2007)	"Priyadarshini Soot Girani	Manufacturing system performance index.	Sigma level improved result in production of premium quality.
6.	Hsu et al. (2007)	A semiconductor	To determine the process capability.	Adjusted process capability and Savings of 3.4 bn $ per annum.
7.	Isaacson (2008)	Cranberry drink and beverages	Packaging process improvement.	DPMO level improved from 3011 to 178 & sigma.
8.	Kumar and Sosnoski (2009)	Wilson Tools and Machines	Waste incurred in Amada A-Station.	Roughly $10,000 per year in savings and Reduce cycle time.

9.	Hajduova et al. (2009)	A copper wire manufacturing	Quality improvement, customer complaints.	Increased customer satisfaction and new clients, Market share.
10.	Mekki (2010)	A Circuit breaker	Customer complaints- Gas & pressure leakage.	Reduced gas & pressure leakage and reduction in defect.
MICRO SCALE UNITS				
1.	Hutchins (2000)	Carriage and Wagon Works	Axle rejection due to cracks along length.	59% reduction in axle rejections. Annual savings.
2.	Ingram (2000)	Tata Honeywell Limited	Travel costs per kilometer per month.	Decreased travel cost. Reduced the maintenance cost by 25%.
3.	Henderson and Evan (2000)	General Electric company	Keys for implementation of quality tools.	Found Keys factors for successful implementation.
4.	Prasad (2002)	A Bulb manufacturing	Shell cracking of bulbs become un-controlled.	Sigma level increased from 3.2 to 4.5 with reduction in wastage
5.	Rasis et al (2002)	Paper organizers international	Metallic securing devices breaking.	DPMO level reduced from 625000 to 6210 and Savings.
6.	Hammer (2002)	Indian automobile	Implementation of quality tools in India.	Finding Success factors, reduce scrap and cycle time by 20%.
7.	Gack and Robison (2003)	Welding Process	Strength and variability in welded joints.	Strength increased by 25%. Weld life improved by 38%.
8.	Sharma (2003)	Baxer Battery	Financial and Customer satisfaction.	Reduced cost of capital and improvement in quality dealing.
9.	Lipscomb and Lewis (2004)	An automotive application	Structural reliability.	A 3.76 sigma level achieved. Maintenance decrease by 30%.
10.	Motwani et al. (2004)	Dow's Chemicals	Culture change and more accidents.	$1.5 billion saved annually. Costs reduced by 20%.

Figure 2.1 demonstrates the growing gap between manufacturing and service focused Six Sigma research over years. Although modifications have been made in the Six Sigma framework to extend its application from manufacturing to service sector, the increasing gap between the numbers of manufacturing and service focused Six Sigma articles since 2005 implies the return of Six Sigma to manufacturing as its initial base.

Figure-2.1 Six Sigma in Manufacturing and Service Sector

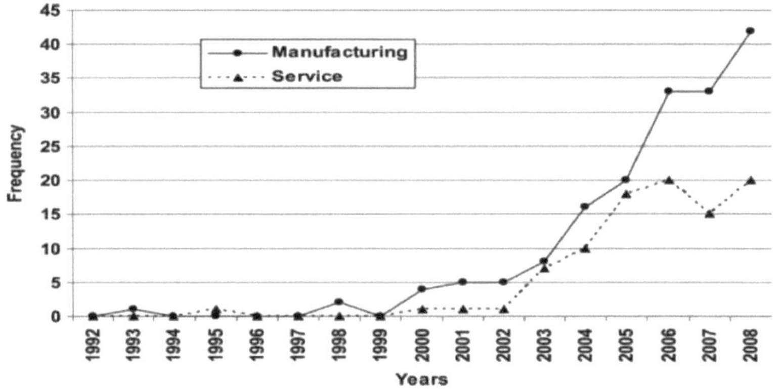

(Source: Aboelmaged, 2009)

2.3 Six Sigma in Indian Industry

For global competitiveness, Indian industries are extensively engaged in Quality Circles, TQM and ISO Certification to achieve overall operational excellence. However, these methods have failed to deliver required performance over the last decade or so (Chaudhery, 2010). It appears that Six Sigma has not been fully explored by Indian industries. In India, 95 percent of the industrial units are in SME sector and contribute about 40 percent of the value addition by the manufacturing sector (Sadagopan et al., 2005). In most of these units productivity levels are alarmingly low and average growth rate of productivity for Indian SMEs has been 4.95% in comparison to 7.31% for China, 9.45% for Singapore and 8.65% for Pakistan (Antony and Banuelas, 2002). Antony and Desai (2009) have found that more than 50 per cent of the companies had applied Six Sigma in their entire operations. 30 per cent of the companies applied Six Sigma in the production area (refer figure 2.2). Other prominent areas where Six Sigma was applied were design and project delivery (8 per cent each). Further the study of Antony and Desai (2009) has highlighted that 46 per cent of the companies had their core processes operating at three sigma quality level, 31 per cent of the companies had their core processes operating at a sigma level between 3 to 4 and 15 per cent of the companies having sigma quality level of their core processes between 4 to 5. It was also observed that 8 per cent never calculated sigma level of its core processes.

Figure-2.2 Six Sigma in Indian Industries

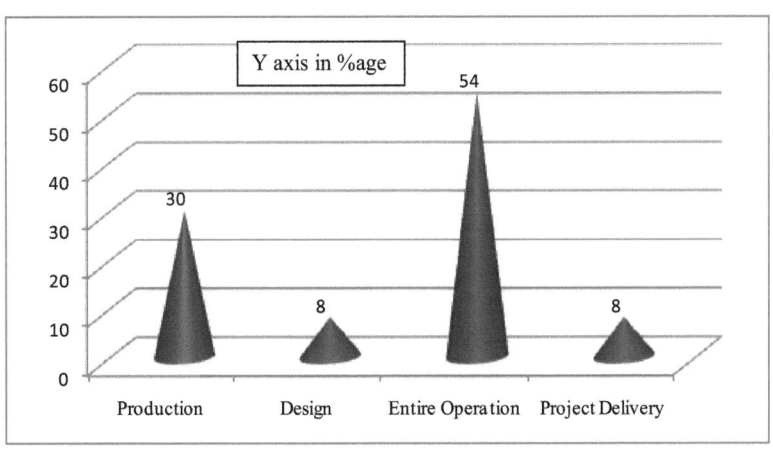

(Source: Antony and Desai, 2009)

Antony (2009), during survey on Indian companies, has formulated some interesting data on the usage, awareness and status of some Six Sigma tools/techniques. As evident from Table 2.5, the most commonly used statistical tools and techniques among the participating companies are Histogram, Run Charts, SPC Charts and Process Capability Analysis. Similarly, the most commonly used problem solving tools are Brainstorming, Cause and Effect Analysis, Pareto Analysis, Process Mapping and Project Charter. The least commonly used statistical tools and techniques are Non-parametric tests (Mann-Whitney Test), Taguchi Methods and Design of Experiments (DOE). The least commonly used problem solving tools are Affinity Diagrams, Force Field Analysis and Matrix Analysis (Singh & Khanduja, 2012b). It was also found that Failure Mode and Effect Analysis (FMEA) is also quite popular in the participating companies whereas QFD was not very popular among the Indian companies (Singh and Khanduja, 2013).

Table-2.5 Awareness on Six Sigma Tools and Techniques in India

Tools/Techniques	Known	Unknown (%)	Usage
Histogram	100	0	4.15
Scatter diagram	85	15	3.41
Run charts	92	8	4.08
SPC charts	100	0	3.92
Cpk analysis	100	0	3.69
MSA	85	15	3.27
DoE	77	23	2.80
Taguchi methods	38	62	3.60
ANOVA	62	38	3.62
Hypothesis testing	62	38	3.50
Regression	77	23	3.10
Flowchart/mapping	100	0	4.10
Brainstorming	100	0	4.30
C and E analysis	100	0	4.30
Affinity diagrams	46	54	2.50
Pareto analysis	100	0	4.08
5S Practice	77	23	3.76
Matrix analysis	38	62	2.62
Balanced scorecard	54	46	3.42
Project charter	92	8	4.00
QFD	85	15	3.09
FMEA	92	8	4.00

Kaizen100	100	0	3.38
SIPOC	77	23	3.70
PDCA	77	23	3.60
Poka-Yoke	77	23	3.50
Benchmarking	77	23	3.80
Quality costing	69	31	3.66

(Source: Antony, 2009)

(Known: Familiarity of given tool, Unknown: Non-familiarity of tool, Usage: Actual utility on 1 to 5 scale)

Six Sigma has evolved into a powerful business improvement methodology in many Indian industries and its importance is growing (Singh & Khanduja, 2011a). During the questionnaire based survey in India by Antony (2009), the respondents were asked to share their experiences on Six Sigma benefits for their organizations since implementation. The respondents were asked to rate their experiences on benefits gained from Six Sigma program on a Likert scale of 1 to 5 (1= no benefit; to 5= excellent benefits). They provided their ratings against different benefit criteria as indicated in table 2.6.

Table-2.6 Perceptions of Indian Companies on Six Sigma Benefits

Benefiting Criteria	Mean score
Reduction of scrap rate	3.64
Reduction of cycle time	3.45
Reduction of delivery time	3.27
Reduction of complaints	3.45
Increase in productivity	3.27
Reduction of process variability	3.63
Reduced need for checking	3.27
Reduction of operational costs	3.81
Increase in profitability	3.54
Improvement in company image	3.00
Improvement in employees morale	2.90
Improved attitude of employees	3.00
Improved attitude of management	3.27

(Source: Banuels and Antony, 2003)

The results show that most Indian companies know the various benefits incurred from Six Sigma approach but still around 30 per cent of them have applied Six Sigma in their business operations and the remaining 70 per cent are yet to experience the Six Sigma initiative for a

number of reasons (Panchal, 2010). During literature survey some of the impeding factors in implementation of Six Sigma were highlighted and these are:
- Lack of resources (this includes financial resources, human resources, time etc.)
- Internal resistance (especially political resistance and technical resistance)
- Poor project selection methodology (i.e. there is no structured approach)
- Lack of tangible results and Poor training/coaching.

2.4 Six Sigma in Foundry Sector

The technical and organizational progress of the last years and the always increasing competition has forced the foundry units to improve their activity by adopting modern tools of problem diagnosis and process improvement (Singh & Khanduja, 2012a). Given the specific market realities especially in the area of medium and large castings, the road to improvement is becoming almost mandatory for the business to survive against competition. Among these diagnostic and optimizing methods, the Six Sigma instruments have shown their worth, either used at the macro level of the productive unit or locally, where the need arises. Among the manufacturing industries, the foundry industry in modern civilization may be defined as the mother of the industrial world (Rao and Rao, 2007) There are about 3, 50,000 foundries in the world with annual production of 90 million metric tons, providing employment to about 20 million people (Choudhary, 2010). The Indian foundry industry is the fourth largest in the world. There are more than 70,000 foundries in India, and most foundries (nearly 95%) in India fall under the small and medium scale category (Sadagopan et al., 2005). These units produce a wide range of castings that include automobile parts, agricultural implements, machine tools, diesel engine components etc. Foundries have limitations when employing the Six Sigma programme due to the limited time, cost and effort and other barriers from foundry characteristics (e.g. organisation culture, structure and strategies). Some challenges being faced by foundry SMEs in modern competitive world, have been listed in table 2.7. Foundries need a methodology to guide and facilitate the implementation of Six Sigma, particularly in a systematic way, which might not be the full system of Six Sigma (Dua et al., 2013). To overcome the high costs of consultancy and training, which is a major component of cost during Six Sigma implementation, self-learning is an alternative solution for the foundries.

Table-2.7 Challenges and Advantages for Foundry SMEs

Challenges	Advantages
Ignorance about the Six Sigma and strategic gains. Lack of time and resources for implementing the drive. A misconception like Six Sigma involves lot of statistics which is beyond the understanding of common work force and it is being a luxury sold out at hefty prices (Singh, 2011).	Ease of training, locating and arriving at the consensus. Complete and direct involvement of top management. Convenient to keep a close watch on the business process and give freedom to-do experiments by varying critical process parameters (Singh and Bakshi, 2013).
Generally foundry SMEs are comfortable with tradition of resorting to quick-fix solutions and curing the problem, as when it comes up. It implies usually foundries are indifferent about investing time and money in the long term, permanent and strategic solutions (Singh and Khanduja, 2009).	Foundry SMEs can respond faster to any change. The small size brings benefits of high speed and more leanness to the system. Easy to hear the customer complaints and critical to quality factors can be found easily for necessary action (Singh and Khanduja, 2010f).
Focus is mostly on the operational matters and not on the improvement. No training budget is fixed in foundry SMEs. Commonly looking for low wages labour. No incentive or reward programmes causes demotivation among the work force at all levels (Singh et al., 2011).	Fewer departmental interfaces are there, so top management can lead the whole show more efficiently. Training cost is less and can be held in more focused way. Absence of bureaucracy in SMEs can generate more loyal human resource (Singh and Sodhi, 2014).
Adamant and dictorial nature of few management people or the owner can ruin the whole industry very rapidly. Loose and informal working environment and relationships. Lack of standardization. Lots of myths regarding Six Sigma and their implementations in foundry SMEs (discuss ahead) are prevailing in developing countries like India (Sachin et al., 2013).	Can reap rapid benefits by responding to customer and market fastely. Abrupt decision making is possible. Direct interaction with shop floor people is feasible in SMEs which can remove all the negative myths prevailing about Six Sigma right from their roots (Singh et al., 2013)

(Source: Desai, 2006)

There is no industrial revolution without castings enterprise! Today it is impossible to imagine a railway line, diesel engine, airplane, car, bus, tractor, power, utensils, houses, bridges, express ways etc. without cast and machined components (Carnell and Lambert, 2000). USA tops the list in alloy production followed by China, Japan, Germany and the CIS. In terms of number of foundries, China has the highest number (93740) followed by India (70,000). The outputs (yield) of foundries are measured in 'metric tons of castings' casted per annum. China is leading in this respect as 33.5 million metric tons of castings itself are being casted in China per annum (Mannual, 2006).The second number is US with 10.8million metric ton per annum (MTPA) and third is Russia having yield of 7.8 million MTPA. India is at number four with yield of 7.1 million MTPA (Panchal, 2010). The overall ranking of top ten countries of world with respect to production yield has been shown in figure 2.3.

Figure-2.3 Top Ten Countries in Foundry Production

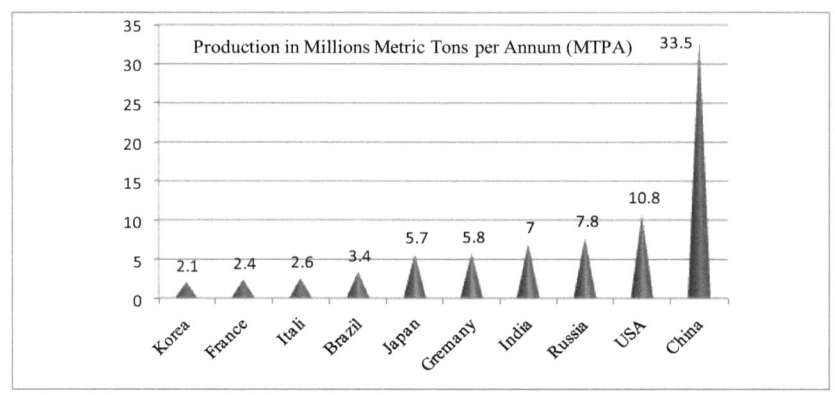

(Source: Panchal, 2010, Indian Foundry Journal)

Among various foundries, the share of iron foundries is the maximum i.e. almost 56.21%, followed by steel with 14.31% and then the non-ferrous ones with 29.48%. The global market potential of metal is estimated to be about USD 30 billion (Choudhary and Jain, 2010). The growing environmental concerns and globalization of economies have led to closure of some 8000 foundries in Europe. These countries have been contemplating to shift their business to the low labour cost centers in developing countries like India (Andersson et al., 2006).

For global competitiveness, many techniques, such as Quality Circles, TQM, ISO Certifications, etc. are being tried. But still, the focus remains on specific problem solving. The need of the hour is to strike global optima and not to waste time, money and energy in finding local optima. The foundry sector needs a breakthrough strategy, which can have multidirectional benefits in shorter duration. Six Sigma has already emerged as one of the most effective business strategies in the large organisations, worldwide. Foundry industries are inherently capable of adopting Six Sigma as breakthrough strategy but they need to be shown the roadmap. The multiple gains achieved by this initial effort of Six Sigma on one of the problems of the company are attractive enough for them to deploy Six Sigma company-wide. Project by project application of Six Sigma in foundries can strengthen their understanding about this strategy along with consolidating on the gains from it. Six Sigma among the small foundries is a much-awaited movement, which can strengthen their bottom lines vis-à-vis contributes in uplifting global economy.

The Indian metal casting industry is as old as the Indian civilization and its primordial manifestations were found in the religious figures like 'Natraja', the dancing deity (Rao and Rao, 2007). The root of modern metal casting industry was laid out in the year 1850 A.D and grew with the development of the engineering sector (Sadagopan et al., 2005). The impetus for foundry sector in India was given by the Jute industry in Bengal and the cotton industry in Mumbai in late 19th century. The establishment of TISCO (Tata Iron and Steel Co. Ltd.), Bengal Iron Company and the IISCO (Steel Plant of Steel Authority of India Limited) led to some remarkable new uses of castings, in domestic as well as industrial areas (Bayle et al., 2001). 32.4% of total production of Indian foundries goes to serve only automobile sector, which is quite substantial as compared to other sectors (Chaudhery, 2010). Figure 2.4 represents the output of bifurcated Indian foundries in metric tons per annum (MTPA) and it shows continuous rise in total production of Indian foundries from 2.7 million MTPA (1997-98) to 7.7 million MTPA (2007-08). Due to recent recession, production decreased in 2008-09 to 6.8 million MTPA and now it has again increased to 7.1 million MTPA in 2009-10 (refer figure 2.4).

Figure-2.4 Production Trend of Various Indian Foundries

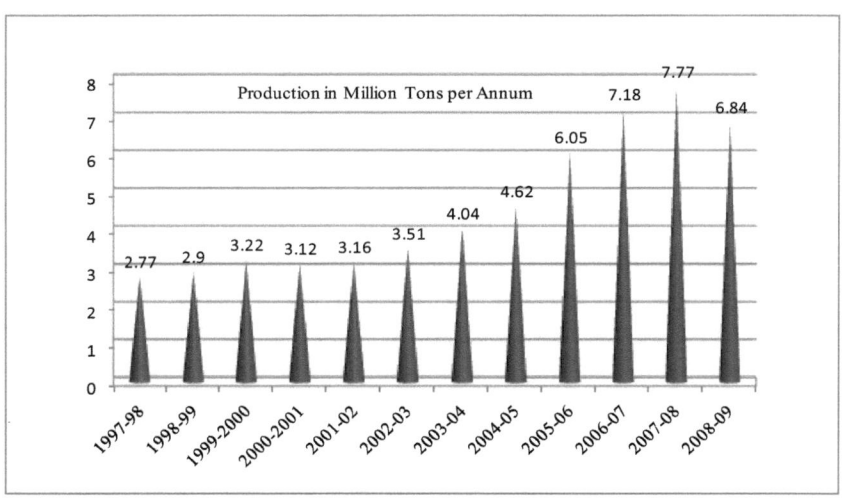

(Source: Chaudhery, 2010, Indian Foundry Journal)

According to the location, the whole country is divided into six major foundry clusters: Belgaum cluster, Batala-Jallandhar-Ludhiana cluster, Delhi-Samalkha-Agra, Coimbatore-Chennai, Kolhapur-Nagpur-Sangli and Rajkot-Ahmedabad cluster. In each cluster there are around 100 to 600 foundry units, which are working under different number of shifts (refer table 2.8). The various types of castings which are produced are ferrous, non ferrous, aluminums alloy, graded cast iron, ductile iron, steel, etc. for application in automobiles, railways, pumps, compressors and valves, diesel Engines, cement, electrical, textile machinery, sanitary pipes, fittings etc. However grey iron castings have a share of about 70% of total production of country. There are around 70,000 units out of which 80% can be classified as small scale units and 10% each as medium and large scale units (refer figure 2.5). Approx 500 units are having 'International Quality Accreditation'. The large foundries are modern and globally competitive and are working at nearly full capacity. Most foundries use cupola furnaces and there is growing awareness on environment with many foundries are switching over to induction furnaces.

Figure-2.5 Bifurcation of Indian Foundries

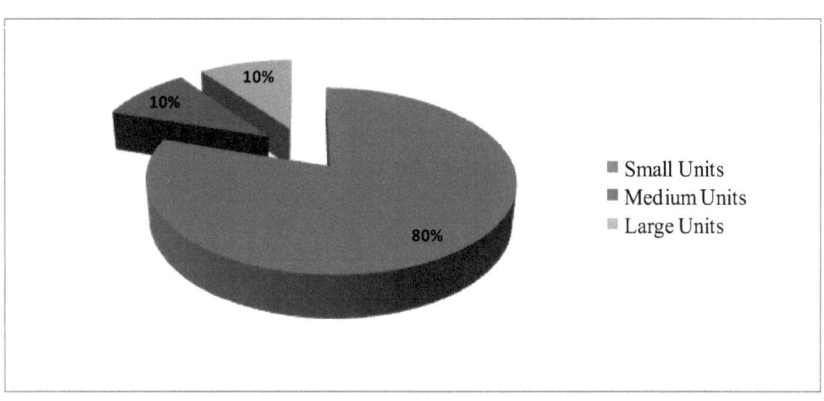

Table-2.8 Details of Major Foundry Clusters

Sr. No.	Foundry Cluster	Number of Units	State	Major Products
1	Batala	250	Punjab	Agriculture implements, machine tools
2	Belgaum	155	Karnataka	Automobile/oil engines, Electric motors
3	Coimbatore	524	Tamil Nadu	Pumps/Valves, Textile machine parts
4	Howarh	375	West Bengal	Machine covers, Sanitary pipes

5	Jalandhar	125	Punjab	Agriculture implements, machine tools
6	Kohlapur	275	Maharashtra	Automobile oil engines, Sugar mill parts
7	Ludhiana	378	Punjab	Sewing machine parts, Auto parts
8	Rajkot	555	Gujarat	Oil engines, Textile machine parts, pumps

The Indian foundry industry is trying to focus on higher value added castings to beat the competition. The total production of Indian foundries is being consumed by 14 different industries or sectors as shown in figure 2.6. It is evident from figure that 32.36% of total production of Indian foundries is consumed by only automobile sector, which is quite substantial as compared to other sectors. Indian foundries can further be classified in five major categories namely; Grey iron foundries, Ductile iron foundries, Steel foundries, Non-ferrous foundries and other miscellaneous foundries.

Figure-2.6 Major Industrial Customers of Indian Foundries

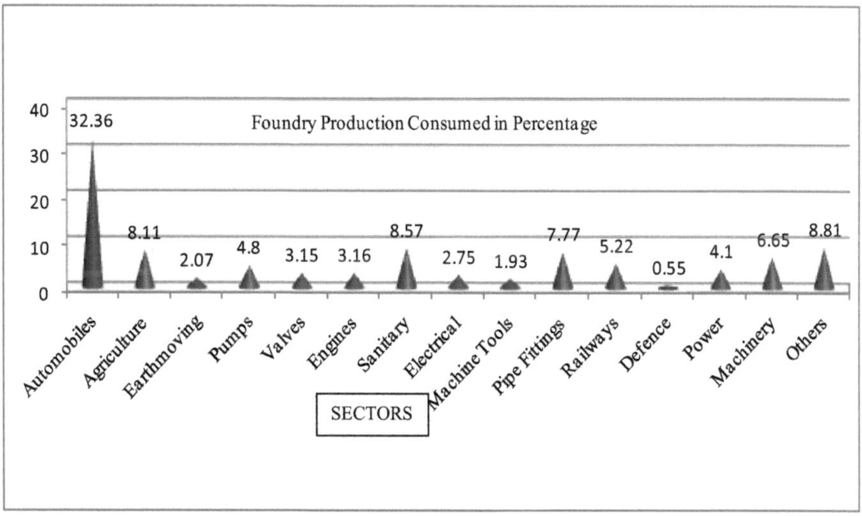

Sector	Foundry Production Consumed in Percentage
Automobiles	32.36
Agriculture	8.11
Earthmoving	2.07
Pumps	4.8
Valves	3.15
Engines	3.16
Sanitary	8.57
Electrical	2.75
Machine Tools	1.93
Pipe Fittings	7.77
Railways	5.22
Defence	0.55
Power	4.1
Machinery	6.65
Others	8.81

(Source: Panchal, 2010, Indian Foundry Journal)

The respective share of each type of foundry in total foundry production is shown in figure 2.7. With the reduction of geographical barriers and the pressure of competing in the global market, overall operational and service excellence has become a necessity for the Indian

Figure-2.7 Classification of Indian Foundries

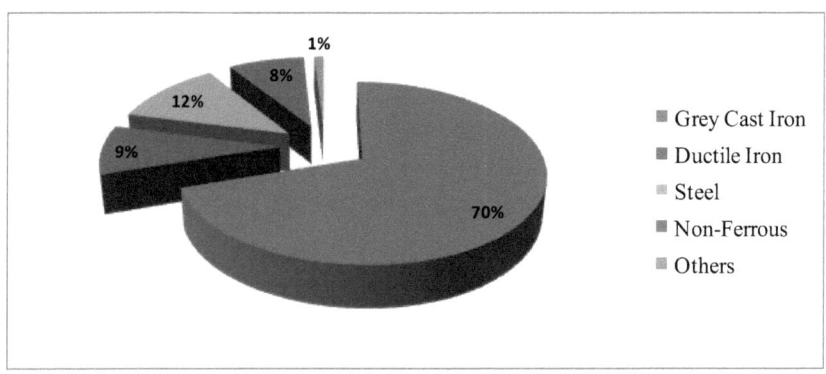

industries to remain globally competitive. Although many Indian industries have successfully embraced the Six Sigma business improvement strategy, the adoption of Six Sigma in Indian industries is not as encouraging as it should be. Approximately 30 per cent of Indian foundries have applied Six Sigma in its business and the remaining 70 per cent of foundries are not yet engaged with a Six Sigma initiative for a number of reasons (Antony and Banuelas, 2002). Satisfaction with other quality and productivity improvement initiatives turned out to be the strongest reason for not embarking on Six Sigma program, followed by lack of awareness and unsuitability of the initiative to their type of business (refer figure 2.9).

Figure-2.8 Reasons for Avoiding Six Sigma in Indian

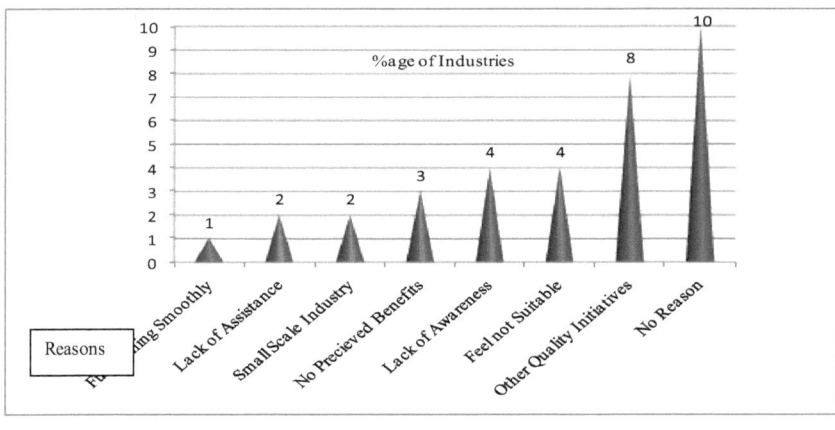

(Source: Antony and Desai, 2009)

36

Due to scarcities of knowledge in the field of Six Sigma, some myths exist in the minds of Indian foundries and these are:

- Six Sigma is the flavor of the month. Senapati (2004) perceives Six Sigma as a fad with the same tools as employed in many other quality initiatives like; Total Quality Management etc. Pojasek (2003) views Six Sigma as another repackaged quality trend that will come and go and is of no help to companies. The author considers Six Sigma as an expensive distraction that requires paying a consultant to walk into an organisation and teach a selected number of people "the newest best way" of problem solving.
- Six Sigma is only for manufacturing companies. Six Sigma originated in Motorola in mid 1980's and was promoted by manufacturing giants like General Electric (GE) and Allied Signal, giving an impression that it can be deployed only in manufacturing companies. The most common reason service-oriented organisations stay away from Six Sigma is that they see it as a manufacturing solution.
- Six Sigma works only in large organisations. It is believed that its application is restricted to large organisations only because of their endless resources and large teams. Small companies might have a more difficult time effectively implementing Six Sigma (Woodall, 2001). Although Six Sigma has been implemented with success in many large corporations, there is still less documented evidence of its implementation in smaller organisations.
- Six Sigma requires strong infrastructure and massive training. Deploying Six Sigma in an organisation requires new skills and this primarily means training the Black Belts and Green Belts who will guide and manage the improvement projects and programs. Employees in the small businesses and public sectors are of the opinion that Six Sigma demands massive training costs and additional effort (Smith and Phadke, 2005).
- Six Sigma is not cost-effective. This is another common myth prevalent in the industrial world. It is presumed that deploying Six Sigma requires massive investment with meager profit or return on investment (ROI). Critics are of the opinion that there are huge risks in heavy investment in this business strategy as it takes a long haul before reaping any tangible benefits (Senapati, 2004).

2.5 Summary

After an extensive literature review, it appears that the application of Six Sigma in the areas other than the big multinational companies has been rare and inconsistent. Little evidence exists in literature on Six Sigma implementation in foundry industry, which is counted as the most important supplier/feeder to other manufacturing industries. Since this industry is facing serious sickness levels, an attempt can be made to implement Six Sigma over various systems and sub systems to considerably improve the productivity and profitability levels.

Literature also reports tremendous financial gains through Six Sigma in large manufacturing sector, so an attempt can be made to replicate it in foundry SME sector too. Due to low level of productivity index, Indian foundries are not keeping up pace with corresponding downstream industries due to which customer companies (like automobile etc.) have to satisfy their monthly needs by importing casted parts from countries like China, USA, Russia, Korea etc. To cope up these challenges, massive micro level implementation of Six Sigma seems to be a highly feasible solution. Moreover, case study approach seems to be dominantly applied from last two decades and reap huge financial and non financial benefits to big or small units. From historical evidences it is obvious that Six Sigma has successfully demonstrated its more compatibility with manufacturing sector as compare to service or process oriented sectors.

Up to now, mostly Indian foundries had more bent towards Japanese techniques to improve quality, which got partial success to meet its objectives due to some inherent professional and social reasons. With increasing application and versatility of Six Sigma abroad, it is now being realized to check its vulnerability in Indian foundry environments, particularly in Indian SMEs.

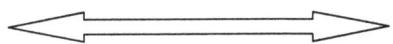

CHAPTER – 3   RESEARCH DESIGN

3.1 Problem Formulation

Besides low level of productivity, Indian foundries have been facing multiple problems like:

- Higher cost of inputs such as; power, raw materials etc.
- Higher import duties on raw materials like; pig iron, cast iron etc.
- Lower productivity due to manual operations and low level of mechanization.
- Non-availability of modern manufacturing techniques, due to high capital cost.
- Non-availability of capital at competitive cost as available in other countries.
- Poor and erratic power availability and low energy efficiency of SME units.
- Lack of foundry specific training facilities and hence scarcity of skilled manpower.
- Cumbersome environmental clearance procedures and foundry pollution.
- Low level of automation
- Focused generally on domestic markets
- Lack of Government Supportive policies
- Un-organized clusterification
- Lack of knowledge and poor information systems.

Many of these problems are of serious concern to production engineers. These can be effectively tackled by using operation management strategies and in this context Six Sigma has emerged as an important diagnosis for these maladies in Indian foundries. So to test the efficacy of Six Sigma and compute its benefits in real terms, a case study was selected to monitor these research imperatives. Case study was carried out at a medium scale foundry unit which had substantially high scrap rate (hence low productivity level) despite presence of apparently high profitability levels. The research plan was formulated in terms of a project approach, where step by step efforts were put in, analyzed and evaluated.

3.2 Book Plan

Objectives of the present study have been chalked out by not ignoring the 'Status-Quo' of Indian foundries, as far as their feeble production metrics are concerned. During vast literature survey conducted to further explore the low productivity levels in foundries, Six Sigma appeared as the most feasible solution. A consistent research plan was formulated to implement a project based approach for Six Sigma in the selected non-ferrous piston foundry. Proper result appraisal was done after executing each phase of DMAIC methodology as per the guidelines of proposed frameworks. Lastly broad conclusions were drawn out and this help to identify the scope for future work. The research plan for the present work has been shown in figure 3.1.

Figure-3.1: Research Plan

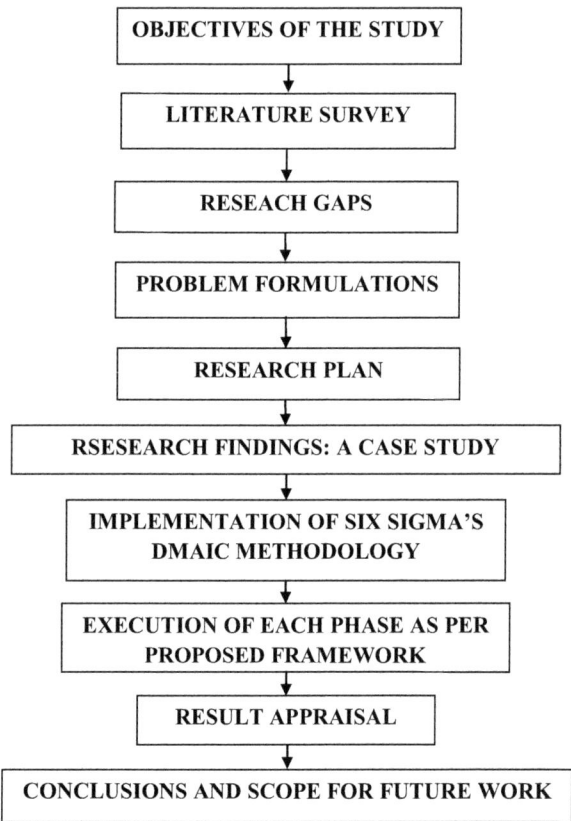

3.3 Methodology Adopted

Six Sigma is a highly structured program and is being used to improve quality all over the world (Velazquez et al., 2010). This approach utilizes number of management/statistical tools and techniques in its respective phases (Lin et al., 2008). There always remains a risk of choosing wrong tools due to negligence or production constraints, which ultimately leads to failure of this approach and it only bounds to produce paper work projects that are far away from real world savings. The proposed work tries to simplify the phases of Six Sigma and categorizes the given tools/techniques with respect to their utility. The work further validates its effectiveness by conducting a successful case study in a non-ferrous foundry. There are five steps in the conventional Six Sigma process:

Define Phase: It is the first phase in the DMAIC model. During this phase, the project's definition is developed (Antony and Banuelas, 2002). The project's definition includes the overall scope, objectives and goals of the project. It also determines the project leader, team members, sponsor, stakeholders and schedule. This is accomplished by utilizing various tools such as Flow Charts, Process Mapping and SIPOC Diagrams. By skipping this phase, practitioners miss an opportunity for making valuable process discoveries. A project that starts directly at the 'Measure Phase' has the potential to be ditched as investigation of the current problem may open a can of worms that diverts attention from the original goal (Markarian, 2004). Usually a significant amount of head scratching is done before practitioners can unveil the right input factors and have a clear understanding of their impacts on the output factor (Yang, 2004). This is why the define stage in the DMAIC model of Six Sigma is so important. This Phase may involve following qualitative and quantitative tools:

- Project Charter
- Voice of Customer (QFD)
- Benchmarking
- Process Mapping (SIPOC, VSM and Process Flow Charts)
- Gap Analysis
- Gantt Charts
- Tree Diagrams

Measure Phase: The dictionary meaning of measure is number or quantity that records a directly observable value or performance (Antony et al., 2004a). This is a critical phase whose goal is to get as much information as possible on the current process so as to fully understand both how it works and how well it works (Kovach, 2003). Measurement system analysis, capability studies and finding performance gaps dominate the work in this phase.

There are number of quantitative and qualitative tools/techniques which can be used for measuring operations and some of these are:
- Process Capability Assessment
- Pareto Charts
- Cause and Effect Diagrams
- Scatter Diagrams
- XY Matrix
- 5Ms
- Matrix Plot
- Activity Network Diagram
- Affinity Diagram
- Prioritization Matrix Grid
- Process Decision Charts
- MSA (Bias Checking, Linearity Measurement, Stability and Gauge R&R etc.)

Analyse Phase: It refers to an examination of processes, facts and data to understand the root cause of given problem. It also determines the various opportunities to bring improvements (Kuei et al., 2003). This stage might include flow charting of process, brainstorming among project team members, satisfying and charting data or conducting formal experiments and statistical tests. This phase tries to uncover the relationship between the response variable (Y) and input variables (Xs). Statistics in particular provides many useful tools to facilitate data analysis (Hutchins, 2000). Measure phase highlights number of critical to quality variables (Xs), but in actual few of them are feasibly responsible and this phase has an objective to focus on those responsible factors, so that breakthroughs can be achieved positively during next phase. Some tools commonly used in this phase are:
- Graphical Data Analysis
- Box Plot Analysis
- Correlation and Regression Analysis
- Process Modeling and Simulation
- Hypothesis Testing (t-Test, u-Test, z-Test and Paired Comparison Tests etc.)
- Analysis of Variance (ANOVA)

- Chi-square Test
- Why-Why Analysis
- Main Effect Plot
- Multi Vary Chart
- Audio Visual Analysis

Improve phase: This phase is just like a goal of Six Sigma project through which improvements or suggestions have been accelerated to achieve unprecedented performance levels by eliminating errors or defects from critical to customer factors/products or processes (Hoerl, 1998). It is the toughest phase of Six Sigma as it is more of an art then a science. The idea is to fine tune critical Xs to optimize response Y and meet customer expectations (Graves, 2002). The success in this phase may depend upon the creativity, innovation and intelligence of Six Sigma team. Some tools relevant to this phase are:

- Evolutionary Optimization (EVOP)
- One-Factor at a Time Experiments (OFAT)
- Multi-Factor at a Time (MFAT)
- DOE (Full or Fractional Factorial, Mixture and Tanguchi's Experiments)
- Response Surface Methodology (RSM)
- Lean Tools for Improvement (Kaizen, Poka-Yoke, 5S etc.)
- Total Productive Maintenance (TPM)
- Mathematical Modeling
- Guidelines or Recommendations from Experiences

Control Phase: It is the last phase of Six Sigma's DMAIC methodology and contains activities to ensure that the project improvements are sustained by tracking key performance process parameters or measures (Fazzari and Levitt, 2008). It includes monitoring of processes, analysis of results and accordingly taking corrective actions when necessary to maintain the business process in central or stable conditions. It is not only the basis of daily management work at all levels but also provides platform for long term improvements. To execute this phase efficiently, some tools which can be effectively used are:

- Process Capability Re-assessment
- Control Charts (Moving Range, Moving Mean and Range Charts etc.)

- Moving Range Charts
- Charts for Attributes (u- Chart, p- Chart, np or np/n Charts etc.)
- Cum Sum Graphs
- Check Lists and Control Plans
- Audit Schedules and Work Instructions
- TPM and TQM
- Training and Documentation
- Use of Visual Controls and Sirens
- Score Boards
- Patrolling Teams

3.4 Tools Used in Present Case Study

In the present case study, certain tools have been strategically selected and used under the respective five phases of Six Sigma. These tools have been shown in table 3.1.

Table-3.1 Tools Used in DMAIC Phases

DEFINE	MEASURE	ANALYSE	IMPROVE	CONTROL
• Voice of Customer (QFD) • Project Charter • Project Scheduling • Snaps of Problem • COPQ Matrix • SIPOC Diagram	• Cpk Assessment • Sigma Calculator • Pareto Chart • Brain Storming by 5M • Value Stream Mapping • Cause & Effect Matrix • MSA • (Gage R&R) • (Bias Checking) • (Stability Test)	• Chi-Square Test • One Way ANNOVA • 2 Samplet-test • Multi-Regression • Interaction PLot • Ishikawa Diagram • Why-Why Analysis	• Optimization of Process Parameters by Full Factorial DoE • Poka-Yoke Principles • Continuous Improvement through Kaizens • On-Job Training to Create Foundry Skills	• MSA Revised • Cpk & Sigma Re-Calculated • Update Control Plans • Work Instructions • FMEA • Patrolling Teams • Process Indicator Board • 5-S

Brief descriptions of these tools have been given as under:

Quality Function Deployment (QFD): This is a method used to relate the characteristics that are customer specific. QFD is similar to the Matrix Diagram but contains more features. This is a structured approach to define customer needs or requirements and translating them into specific plans to produce products to meet those needs. The 'voice of the customer' is the term to explain these stated and unstated customer needs or requirements. The voice of the customer is captured by variety of ways; direct discussion or interviews, surveys, focus groups, customer specifications, observation, warranty data, field reports etc. This understanding of the customer needs is then summarized in a product planning matrix or "house of quality". These matrices are used to translate higher level "what's" or needs into lower level "how's" - product requirements or technical characteristics to satisfy these needs.

Project Charter: This document summarizes the Six Sigma project and is the basis for the official authorization. It includes details of the project team and the stakeholders, the mission statement, the problem statement, the business need, the scope, the resources and their authorization and at end the target completion date for each phase. It serves as a reference of authority for the future of the project and establishes the authority assigned to the project manager, especially in a matrix management environment.

Gantt Chart: This chart shows the work breakdown against time. The horizontal axis shows the activities and the vertical axis represents the time (in days, weeks, months etc.) Project planning and scheduling has been based upon these charts.

Histograms: This is a graphical method that represents the distribution of values in a data set. The data values are grouped into ranges. In statistics, a histogram is a graphical representation showing a visual impression of the distribution of data. It is an estimate of the probability distribution of a continuous variable and was first introduced by Karl Pearson. A histogram consists of tabular frequencies demonstrated as adjacent rectangles, erected over discrete intervals (bins) with an area equal to the frequency of the observations in the interval. The height of a rectangle is also equal to the frequency density of the interval i.e., the frequency divided by the width of the interval. The total area of the histogram is equal to the number of data.

COPQ Matrix: COPQ stands for 'Cost of Poor Quality' and was popularized by Harrington in 1987. It includes calculations regarding internal failure costs, external failure costs, appraisal costs and prevention costs etc.

SIPOC Diagram: SIPOC stands for 'Supplier Inputs Outputs and Customers'. A SIPOC Diagram is created during the 'Define' phase of a Six Sigma project to ensure that the

Suppliers, Inputs, Process, Outputs and Customers have been identified and are agreed by all the team members. A SIPOC is completed most easily by starting from the right ("Customers") and working towards the Supplier. A SIPOC diagram is a tool used to identify all relevant elements of a process improvement project before work begins. It helps to define a complex project that may not be well scoped and is typically employed at the Measure Phase.

Cpk Assessment: In process improvement efforts, the process capability index or process capability ratio is a statistical measure of process capability. This is a ability of a process to produce output within specification limits. The concept of process capability only holds meaning for processes that are in a state of statistical control. Process capability indices measure how much "natural variation" a process experiences relative to its specification limits and allows different processes to be compared with respect to how well an organization controls them.

Sigma Level Calculator: It is a software based tool which tries to find out the sigma level of given process by using the principle of defects per million opportunities (DPMO). Total production and scrap quantity have been added for a fixed interval of time. Negative values for sigma are meaningless and are shown as zero. This calculation includes by default the standard 1.5 sigma shift for short-term sigma. The sample and population sizes are used to extrapolate the total defects in the population. The population and defect opportunity are used to calculate the total number of opportunities. DPMO is then calculated using this formula [DPMO = (total defects * 1,000,000) / total opportunities]. The DPMO is then converted to the number of standard deviations for the equivalent right-hand tail fraction of the normal distribution. This number of standard deviations is the base process sigma value.

Pareto Charts: This is a type of ordered histogram often used to evaluate the frequency of occurrence of defects. It is called a Pareto Chart because the distribution of defects typically obeys the Pareto Principle i.e. 20% of the defect categories account for 80% of the defects. It is usual to include the grouped results in the bars and a line graph of the cumulative total. A Pareto Chart, named after Vilfredo Pareto is a type of chart that contains both bars and a line graph, where individual values are represented in descending order by bars and the cumulative total is represented by the line. The left vertical axis is the frequency of occurrence but it can alternatively represent cost or another important unit of measure. The right vertical axis is the cumulative percentage of the total number of occurrences, total cost or total of the particular unit of measure. Because the reasons are in decreasing order, the cumulative function is a concave function.

Brainstorming: It is a group creativity technique by which a group tries to find a solution for a specific problem by gathering a list of ideas spontaneously contributed by its members. The term was popularized by Alex Faickney Osborn in 1953 through the book 'Applied Imagination'. Brainstorming has become a popular group technique and has aroused attention in academia. Furthermore, researchers have made modifications or proposed variations of brainstorming in an attempt to improve the productivity of brainstorming. However, there is no empirical evidence to indicate that any variation is more effective than the original technique.

5Ms: In the management literature 5 M's of management called as; Money, Men, Material, Method, Machine and nowadays Marketing is placed on the literature as the 6th M of management. In business it pays to have the right innovations but even when you have innovations you still need a follow through plan. Each of these 5 M's is important to the overall production for innovations. The 5 M's of management work together to supply the product. It is important to have the right manpower, right method, right material, right machine and right amount of money to execute or manage a business activity, efficiently.

Value Stream Mapping: It is a lean manufacturing technique used to analyze and design the flow of materials and information required to bring a product or service to a consumer. It can be applied to nearly any value chain and can be drawn by adopting following method; firstly, draw while on the shop floor a current state value stream map which shows the current steps, delays and information flows required to deliver the target product or service. This may be a production flow (raw materials to consumer) or a design flow (concept to launch). There are 'standard' symbols for representing supply chain entities. Next assess the current state value stream map in terms of creating flow by eliminating waste. Value stream mapping is a supporting method that is often used in lean environments to analyze and design flows at the system level (across multiple processes). Although value stream mapping is often associated with manufacturing, it is also used in logistics, supply chain, service related industries, healthcare, software development and product development. Value stream maps are often drawn by hand in pencil; to keep the mapping process real-time, simple and iterative by allowing for simple correction. However, software tools can also be used.

Cause and Effect Matrix: This is a tool to prioritize potential causes by examining their relationship with the CTQ factors. CTQs are placed on the top of the matrix and causes are placed along the left side. The CTQs are ranked in terms of importance. The relationship between the causes and CTQs are ranked. An overall score is calculated and the cause with the highest overall score should be addressed first because they will have the largest impact on the CTQs. It can be chalked out by given steps; first of all list the CTQs across the top of a matrix and assign scores to each CTQ according to its importance to the customer. Then list the causes on the left side of the matrix and determine correlation scores between each cause and CTQ based on the strength of their relationship.

Measurement System Analysis: It is defined as the analysis of a measuring system to determine the amount of variation in the measurement. The amount of variation in the measuring system is often overlooked, but is a significant component of the observed variation. The types of measurement errors can be categorized into: bias, linearity, stability, repeatability and reproducibility. It is a scientific and objective method of analyzing the validity of a measurement system through monitoring equipment variation, Appraiser (operator) variation and the total variation of a measurement system. Measurement Systems are much more than the measuring instruments and gauges used for measuring. The measurement value is a result of the measurement process carried out by:

- The measuring instrument
- The person using the measuring instrument (Appraiser)
- The environment under which the reading has been obtained
- The methods used and its setup
- The tooling and fixture that locates and orients the object under measurement

Accuracy/Bias: It is the difference between the observed average value and the master reference value. Bias is statistically significant and is a systematic shift of the reading from its true master value. Bias is usually attributed either to an instrument error, that adds (or subtracts) a constant value to each reading. This can be due to a worn out instrument or a parallax like error in the appraiser, who evaluate the reading.

Repeatability and Reproducibility (R&R): Repeatability error is the inherent random variation in the instrument. Reproducibility is the error induced by the influence of the appraiser. R&R errors are usually addressed by providing fixtures for a uniform means of handling the work piece during the measurement process and standardizing the methods of measurement by training the appraisers. Variations induced by environmental effects are also classified under reproducibility. It is important to make the measurement system robust to all the environmental variations that can normally occur during the course of measurement.

Linearity: Measuring instruments are often used at various nominal dimensions along their scales. Linearity evaluates whether the bias is uniform across the operating range of the measurement system.

Stability: Stability study monitors the state of the measurement system over a period of time. A measurement system will induce more variations in the readings due to wear and tear as it gets into use. Each measurement system will go out of stability after different intervals based on their usage. Rather than fixed period schedules, Stability test can be an excellent guideline to signal when a measurement system should be taken up for calibration. Using Stability Study to determine the calibration frequency can lead to bottom-line savings for your organization. Stability Test scientifically assures you of the predictability of the measurement system behavior over an extended time period. Need of training is clearly highlighted by evaluation of appraiser to appraiser variation. The quality of a product that is shipped to the esteemed customer can only be as good as the measurement systems that go into measuring it.

Chi Square Test: A chi-square test (chi squared test or χ^2 test) is a statistical hypothesis test. The chi-square test is used to examine differences with categorical variables. It is used in two similar but distinct circumstances; for estimating how closely an observed distribution matches an expected distribution and for estimating whether two random variables are independent.

One Way ANOVA: It is a technique used to compare means of two or more samples (using the F distribution). This technique can be used only for numerical data. The ANOVA tests the null hypothesis that samples in two or more groups are drawn from the same population. To do this, two estimates are made of the population variance. The ANOVA produces an F statistic, the ratio of the variance calculated among the means to the variance within the samples. If the group means are drawn from the same population, the variance between the group means should be lower than the variance of the samples, following central limit theorem. A higher ratio therefore implies that the samples were drawn from different populations. Typically, however the one-way ANOVA is used to test for differences among at least three groups, since the two-group case can be covered by a t-test (Gosset, 1908). When there are only two means to compare, the t-test and the f-test are used. The results of a one-way ANOVA can be considered reliable as long as the following assumptions is met; Response variable must be normally distributed (or approximately normally distributed) and samples are independent.

2 sample t-test: The two-sample t-test is a hypothesis test for answering questions about the mean when the datas are collected from two random samples of independent observations. It tells about the variation in mean of respective populations. The steps of conducting a two-sample t-test are quite similar to those of the one-sample test. A comparison of this sort is very common in medicine and social science. To evaluate the effects of some intervention, program or treatment, a group of subjects is divided into two groups. The group receiving the treatment to be evaluated is referred to as the treatment group, while those who do not are referred to as the control or comparison group. A hypothesis test used to compare the means of two reasonably small (30 or less) samples to see if it is feasible that they come from the same population.

Multi Regression Analysis: The general purpose of multiple regression (the term was first used by Pearson, 1908) is to learn more about the relationship between several independent or predictor variables and a dependent or criterion variable. A linear regression model relates the response to several inputs (see figure 3.2).

Figure-3.2 Regression Analysis

The steps in multiple regressions are basically the same as in simple regression. State the research hypothesis and the null hypothesis and then gather the data and assess each variable separately. Assess the relationship of each independent variable, one at a time with the dependent variable (calculate the correlation coefficient and obtain a scatter plot). Find the relationships between all of the independent variables with each other (obtain a correlation coefficient matrix for all the independent variables). Then calculate the regression equation from the data and accept or reject the null hypothesis. Finally explain the practical implications of the findings.

Fish Bone Analysis: Ishikawa diagrams (also called fishbone diagrams or herringbone diagrams or Fishikawa) are causal diagrams that show the causes of a certain event. It is created by Kaoru Ishikawa (1990). Common uses of the Ishikawa diagram are product design and quality defect prevention, to identify potential factors causing an overall effect. Each cause or reason for imperfection is a source of variation. Causes are usually grouped into major categories to identify these sources of variation. The categories typically include:

- People: Anyone involved with the process
- Methods: How the process is performed and the specific requirements for doing it, such as policies, procedures, rules, regulations and laws
- Machines: Any equipment, computers, tools etc. required to accomplish the job
- Materials: Raw materials, parts, pens, paper, etc. used to produce the final product
- Measurements: Data generated from the process that are used to evaluate its quality
- Environment: The conditions, such as location, time, temperature, and culture in which the process operates

This is a graphical tool used to list and categorize possible causes of a problem. It looks like a fish skeleton and is sometimes called a 'fishbone diagram'.

Why-Why Analysis: This is a question asking method used to explore the cause/effect relationships underlying a particular problem. Ultimately, the goal of applying the Why-Why method is to determine a root cause of a defect or problem. Why-Why helps to identify how to really prevent the issue from happening again. A Why-Why is most effective in a team setting or with more than one person involved and capture the inputs on a flipchart or a simple spreadsheet like the one below;

- First start with the problem to solve. Then ask, "Why is x taking place?" It will be ended up with a number of answers. Further explore down the each answer.
- Repeat the process for each of the answers to the first question.

- During this iterative process usually some root causes have been highlighted. Now one can identify specific action plans to address those root causes.

Involve the right people– it helps to have those that are familiar with the process and the problem in the room, so they are able to answer why something happened.

Full factorial DOE: It is an experiment whose design consists of two or more factors, each with discrete possible values or "levels" and whose experimental units take on all possible combinations of these levels across all such factors. A full factorial design may also be called a fully crossed design. Such an experiment allows studying the effect of each factor on the response variable, as well as the effects of interactions between factors on the response variable. For the vast majority of factorial experiments, each factor has only two levels. For example, with two factors each taking two levels, a factorial experiment would have four treatment combinations in total and is usually called a *2×2* factorial design. If the number of combinations in a full factorial design is too high to be logistically feasible, a fractional factorial design may be used, in which some of the possible combinations (usually at least half) are omitted. For more than two factors, a 2^k factorial experiment can be recursively designed, where k represents the number of factors.

Kaizen (改善): It is a Japanese word which stands for "improvement" or "change for the better". It refers to philosophy or practices that focus upon continuous improvement of processes in manufacturing, engineering and business management. It has been applied in healthcare, psychotherapy, life-coaching, government, banking and other industries. When used in the business sense and applied to the workplace, kaizen refers to activities that continually improve all functions and involves all employees from the CEO to the assembly line workers. By improving standardized activities and processes, kaizen aims to eliminate waste. Kaizen was first implemented in several Japanese businesses after the Second World War. It has since spread throughout the world and is now being implemented in many other venues besides just business. Kaizen is a daily process, the purpose of which goes beyond simple productivity improvement. It is also a process that when done correctly, humanizes the workplace, eliminates overly hard work ("muri") and teaches people how to perform experiments on their work using the scientific method and how to learn to spot and eliminate waste in business processes.

Poka Yoke (ポカヨク): It is also a Japanese term that means "fail-safing" or "mistake-proofing". A poka-yoke is any mechanism in a lean manufacturing process that helps an equipment operator avoid (*yokeru*) mistakes (*poka*). Its purpose is to eliminate product defects by preventing, correcting or drawing attention to human errors as they occur. The concept was formalised and the term adopted by Shigeo Shingo as part of the Toyota Production System. It was originally described as *baka-yoke*, but as this means "fool-proofing" (or "idiot-proofing") the name was changed to the milder *poka-yoke*. More broadly, the term can refer to any behavior-shaping constraint designed into a product to prevent incorrect operation by the user. Poka-yoke can be implemented at any step of a manufacturing process where something can go wrong or an error can be made.

X bar and R charts: These variable charts are used as a pair to plot the mean and range of subgroups taken from the process. In statistical quality control, the $\bar{x}$ and R chart is a type of control chart used to monitor a variable's data when samples are collected at regular intervals from a business or industrial process. The chart is advantageous in the situations when the sample size is relatively small (say, n $\leq$ 10—$\bar{x}$ and s charts are typically used for larger sample sizes)

p Charts: These are attribute type charts used to plot units nonconforming, when the samples are not of equal size. This is often when samples form a natural grouping - for example the number of treatments in a hospital in a week. Note that the control limits are dynamic, they depend on the size of the sample only.

Control Plan: It is a management tool to identify and monitor the activities required to control the critical inputs or key outputs for a process so the process will continually meet its product or service goals. This Plan is usually supported by a chart which captures the specific measurements of either the process inputs or outputs and is used to evaluate whether the process is in or out of control. The plan itself is a statement of how an organization plans to monitor the performance of a process, the data that needs to be gathered, the frequency of the collection and the appropriate control limits which designates performance that management will accept. It is used in the control phase to identify and record controls, targets and specification limits for Key Process Input Variables and Key Process Output Variables.

Check Sheets: The check sheet is a simple document that is used for collecting data in real-time and at the location where the data is generated. The document is typically a blank form that is designed for the quick, easy and efficient recording of the desired information, which can be either quantitative or qualitative. When the information is quantitative, the check sheet is sometimes called a tally sheet. A defining characteristic of a check sheet is that data is recorded by making marks ("checks") on it. The check sheet is one of the seven basic tools of quality control.

5S: It is the name of a workplace organization methodology that uses a list of five Japanese words which are; *seiri, seiton, seiso, seiketsu* and *shitsuke*. Transliterated or translated into English, they all start with the letter "S". The list describes how to organize a work space for efficiency and effectiveness by identifying and storing the items used, maintaining the area and sustaining the new order. It also instills ownership of the process in each employee.

Exercise

1. What is the significance of Problem Formulation in research work?
2. What do you mean by Project Charter? What is its role in define phase?
3. Give full form of COPQ? Why we generate it during six sigma projects?
4. Demonstrate various informations which we get from SIPOC Diagrams?
5. How Cpk is different from Cp?
6. Discuss importance of 5Ms as far as work space management is concerned?
7. What do you mean by VSM? What kind of information it convey?
8. Define MSA? What is its significance? Discuss at least three elements of MSA?
9. Explain how DoE is better than traditional experimental techniques?
10. Write a short note on FMEA? How one can use it in Control phase of DMAIC Project?

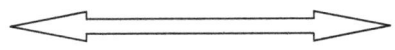

CHAPTER – 4 AN INDIAN CASE STUDY

4.1 Introduction

The study focuses on scrap reduction in foundries and tries to find out the reasons for low productivity index. It also dwells on shattering various myths among SMEs on Six Sigma and its implementation, by validating its compatibility in a case study of an Indian foundry. Doing things right in first time and keeping them consistent is the primary aim behind Six Sigma. An empirical investigation has been carried out in a make-to-order type (medium sized) foundry in which, 'DMAIC' Six Sigma approach has been implemented successfully to decrease the scrap (or defects) of piston castings from 21.2% to 10.4%. Present case study involving Six Sigma implementations in a foundry spanned from 1st July, 2010 to 31st Dec, 2010 and it was a rare research experience in an Indian foundry environment.

4.2 About the Organization

A case study has been carried out in a non-ferrous foundry at Federal Mogul (I) Limited at Bhadurgarh (Patiala), India. The unit casts around 9.5 million pistons annually and has a covered area of about 50144 sq. m. It was established in 1954 and is a medium scale unit casting pistons by semi-automatic die-casting machines. Dies of different types of pistons have been installed on machines as per the monthly planning and pouring of metal is performed manually by operators. These pistons find application in bi-wheelers, cars, SUVs, tractors, light commercial vehicles, heavy commercial vehicles, heavy output diesel engines, stationary engines etc. The piston foundry is used to die cast four different type of pistons as shown in table 4.1.

Table-4.1 Production Capability of Foundry

Piston Types	Capacity (Million/year)
Gasoline (Domestic Market)	7.3
Diesel (Domestic Market)	2.5
Gasoline Cars (Foreign Market)	4.0
Total Production	13.8

Production capacity is around 14 million pistons per annum and product range covers pistons of 30mm to 300mm diameter (refer figure 4.1).

Figure-4.1 End Product (Pistons) Range

The current market share of Federal Mogul (in international and domestic market) has been shown in figure 4.2.

Figure 4.2 Market Share Descriptions

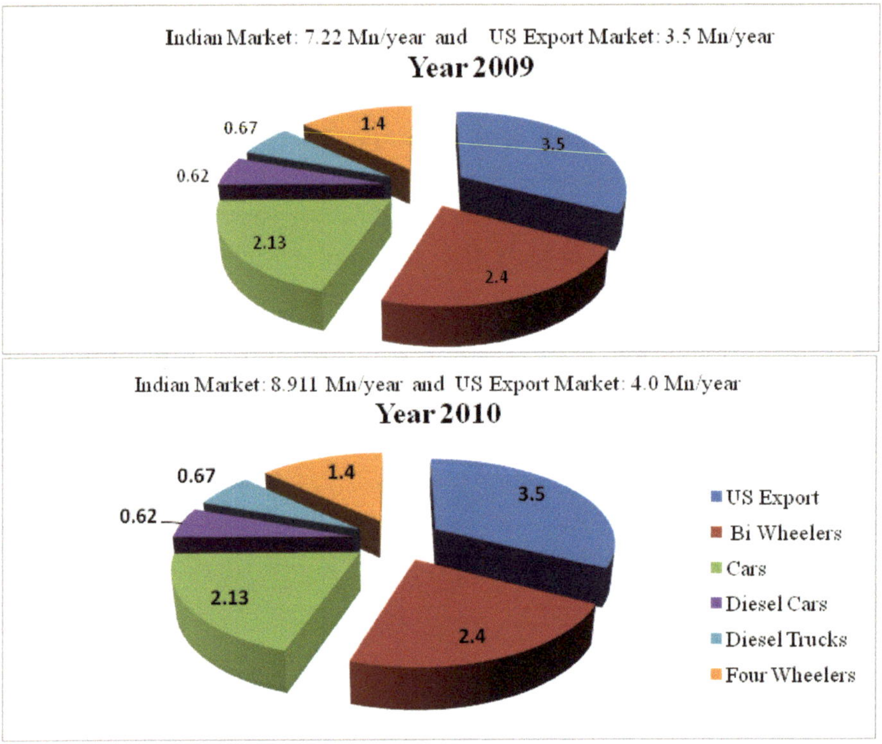

This figure consists of two pie-charts representing market share of Federal Mogul in year 2009 and 2010 respectively. It is evident from the figure that the foundry has not only increased its production of pistons in domestic market (i.e. from 7.22 Mn to 8.91 Mn) but it has also raised its share in foreign market (i.e. from 3.5Mn to 4.0 Mn) from year 2009 to 2010. Further bifurcation of domestic market has been done in terms of pistons for; bi-wheelers, cars, four wheelers, diesel cars and diesel trucks.

4.3 Implementation of DMAIC Approach

In India, manufacturing industries like foundries do not enjoy monopoly and they have to face tough competition (Antony, 2004b). To retain the share of the market, it is necessary to constantly improve the quality of the cast product without increasing the price. The price is influenced by the cost of production, which in turn is influenced by rework or rejection rate. Attention on quality assurance can reduce the wasteful rework and can reduce the cost of casting production. This quality production results in the company's growth and profitability. In a small unit, where investment in plant and machinery is less than Rs.10 million, productivity and profitability are indispensable to assess the performance of such an organization. Black and McGlashan (2006) have observed that reduction in cost and product rejection rate are among the main pressures on small units. The main barriers for these units to be competitive are inadequate technologies causing lot of defects, poor human expertise and scarcity of resources. These units have been looking for some compatible quality initiatives that enhance capability to uplift their business processes for long term survival. Embarking on a Six Sigma programme means delivering top-quality service and products while virtually eliminating all internal inefficiencies. A true Six Sigma organization produces not only excellent product but also maintains highly efficient production and administrative systems that work effectively with the company's other service processes (Henderson and Evans, 2008). In administrative processes, Six Sigma may mean not only the obvious reduction of cycle time during production but more importantly, optimizing response time to inquiries, maximizing the speed and accuracy with which inventory and materials are supplied and fool proofing such support processes from errors, inaccuracies and inefficiency (Thawani, 2004). Much can be achieved from programmes like Six Sigma with the active, consistent, innovative, continuous and widely apparent participation by top management. The primary factor in the successful implementation of a Six Sigma project is to have the necessary resources, the support and leadership of top management, customer requirements identified explicitly and a comprehensive training programme (Caulcutt, 2001). At the heart of the Six Sigma methodology is the DMAIC (define, measure, analyze, improve, and control) model for existing processes that fall below specifications and seek incremental improvement. DMAIC (pronounced "Duh-MAY-ick") is a structured problem-solving methodology widely used in diversified businesses. In brief the DMAIC project methodology has five phases (Lee and Choi, 2006):

- *Define* the problem, the voice of the customer and the project goals, specifically.
- *Measure* key aspects of the current process and collect relevant data.
- *Analyze* the data to investigate and verify cause-and-effect relationships. Determine what the relationships are and attempt to ensure that all factors have been considered. Seek out root cause of the defect under investigation (Lanyon, 2003).
- *Improve* or optimize the current process based upon data analysis using techniques such as design of experiments, poka-yoke or mistake proofing and standard work to create a new future state process. Set up pilot runs to establish process capability.
- *Control* the future state process to ensure that any deviations from target are corrected before they result in defects (Zhan, 2008).

Present book defines the Six Sigma opportunity as a guide for choosing those processes that will reap the greatest corporate benefit from Six Sigma projects. Besides being given guidelines for establishing an improvement opportunity, specific methods for capturing the voice of the customer and ways to use Six Sigma improvement opportunities from a financial perspectives, have also been explored. This work has also proposed conceptual frameworks for successful implementation of each phase separately and some imperatives for right implementation of Six Sigma, by not ignoring actual production constraints of Indian foundries, have also been dwelled in the research.

4.4 Define Phase

This phase identifies suitable projects based on business objectives, customer needs and feedbacks. Critical to Quality (CTQ) issues and other items having an impact on quality and customer satisfaction have also been identified. Six Sigma projects need to be selected mindfully and scoped appropriately. Yet, without known solutions identifying projects that

will truly improving the bottom line can be challenging (Singh & Khanduja, 2012a). The ordinary tasks in this phase are:
- Select project with a significant impact on business.
- Create and validate business case that expresses the value of the project in quantified financial or strategic terms.
- Communicate clearly to the organization with strategic view and the need for change.
- Determine necessary metrics to measure the problem and the success of the project.
- Set the project goals based on business needs and client needs.
- Ensure team member's availability and include the process owner as a part of the team, from the beginning (Yeung, 2007).

The Define phase should be more business-centered than process-centered (Zu et al., 2008). The following tools have been used to determine the business value in the case study:
- A 'project charter' to confirm quantified goals and expected results besides the availability of necessary input resources.
- The 'voice of customer' to understand the business needs from all important internal and external customers of the project.

In past, more Six Sigma deployment projects followed something that might be termed the MAIC (Measure, Analyze, Improve, Control) methodology, but not DMAIC. The goals are based on what was important to the customer and also to business (Green, 2006). This definition method results in more involvement of leadership. Many organizations have taken DMAIC to mean that the Black Belts should be responsible for selecting and defining the projects. This delegation of project definition may have to face serious consequences, because of:
- ***The wrong choice of project.*** This means a project that does not have significant impact to the business or is not aligned to the business strategy. Sometime projects are chosen because they are the 'pet' project, even though they are not truly significant to the business. A better approach is to have strategic leadership and determine the project based on rankings of importance to the customer and to the business (Holtz and Campbell, 2004).

- *The wrong focus.* Again, this situation requires leadership to create the focus based on the short- and long-term goals of the organization.
- *Project takes too long.* When the project is not strategically important, leadership may allow the Black Belt to set the timelines, cause the project to lag or it may be difficult to acquire the proper resources, when the project is not a priority.
- *Projects have insignificant results.* There may be results, but maybe it took too long to achieve them, or because the project was not scoped or chosen properly, the results are not significant to the organization's strategic goals. Having leadership involvement and ownership helps to prevent these problems. It also sends a clear message to the organization that Six Sigma activities are important to the business. This is critical when trying to change a culture (Hare, 2005).
- *Wrong deployment plan.* The culture of a company should provide an environment and foundation to encourage problem solving, excellence and continuous improvement through Six Sigma deployment. The key to designing a foundation is to know or anticipate the needs of the organization. The organizational assessment helps in scripting a deployment strategy and unfurling the underlying culture. If the organizations culture is not suitable to Six Sigma, the culture must be changed before a Six Sigma process is introduced. The Six Sigma requires support from the top brass of the organization. It includes restructuring of the organization to provide supporting infrastructure, training, rewards and communications.
- *Inactive participation of the senior executives.* Six Sigma programme will not survive for a long period without support from senior executives. The senior executives should provide leadership and create a vision, provide necessary resources, give time and review the progress. He should take the responsibility to ensure that everyone gets involved in the implementation effort (Graves, 2002).
- *Lack of technical support.* Often the Black Belts need support from senior executives or Champions on organizational issues and technical issues. The Master Black Belt (MBB), who is teacher and coach, provides all the required technical support to Black Belts (BB's).

- *Scarcity of training.* The first step towards implementation of Six Sigma is building a team of professionals who are trained with varying levels of proficiency in the art of Six Sigma.

After realizing the increasing significance of Define phase or problem defining criteria in industries and services, a simpler and straightforward conceptual frame-work has been proposed for the direct use of Six Sigma practitioners.

4.4.1 Proposed Framework for D-Phase

Today, some companies are still omitting this phase because a project's definition is seen as mainly management work. Following this logic, the same might be true for control: ensuring that the process improvements are implemented and monitoring is only a business responsibility. It has been surveyed from the literature that 35% DMAIC projects fail just because of wrong tool selection in define phase, as per the given conditions and business constraints (Furterer and Elshennawy, 2005). An appropriate define-tool or technique can only give true results when applied to compatible business scenarios at right time. To cope up with this challenge, present book has formulated a frame work (as shown in figure 4.3) which is basically a tool selection criteria and provide necessary guidelines to short list various define tools for right place and on right time. Basically, with this proposed approach the task of 'Define' is completed in three steps. During first step, aim of define is divided into two sub-aims that is;

- Identify the problem and convert it into project
- And focus on the selected project.

In the second step, each sub-aim is bifurcated into four simple questions respectively, which must be answered for successful completion of overall aim of define phase. Actually these questions are some checks which must be performed by some suitable define tools and are explained in third step of frame work. In this way all the tools and techniques have been categorized with respect to their functions. Now objective of the tool and its proper selection as per constraints has become easier and hence there is a less scope left for wrong selection of define-tools/techniques. This frame work is just like a guide liner for define phase and has been validated through a business-case in a medium scale foundry. Six Sigma is about problem solving and process improvement but more importantly, it is about changing the

Figure-4.3 Proposed Frame Work for D-Phase

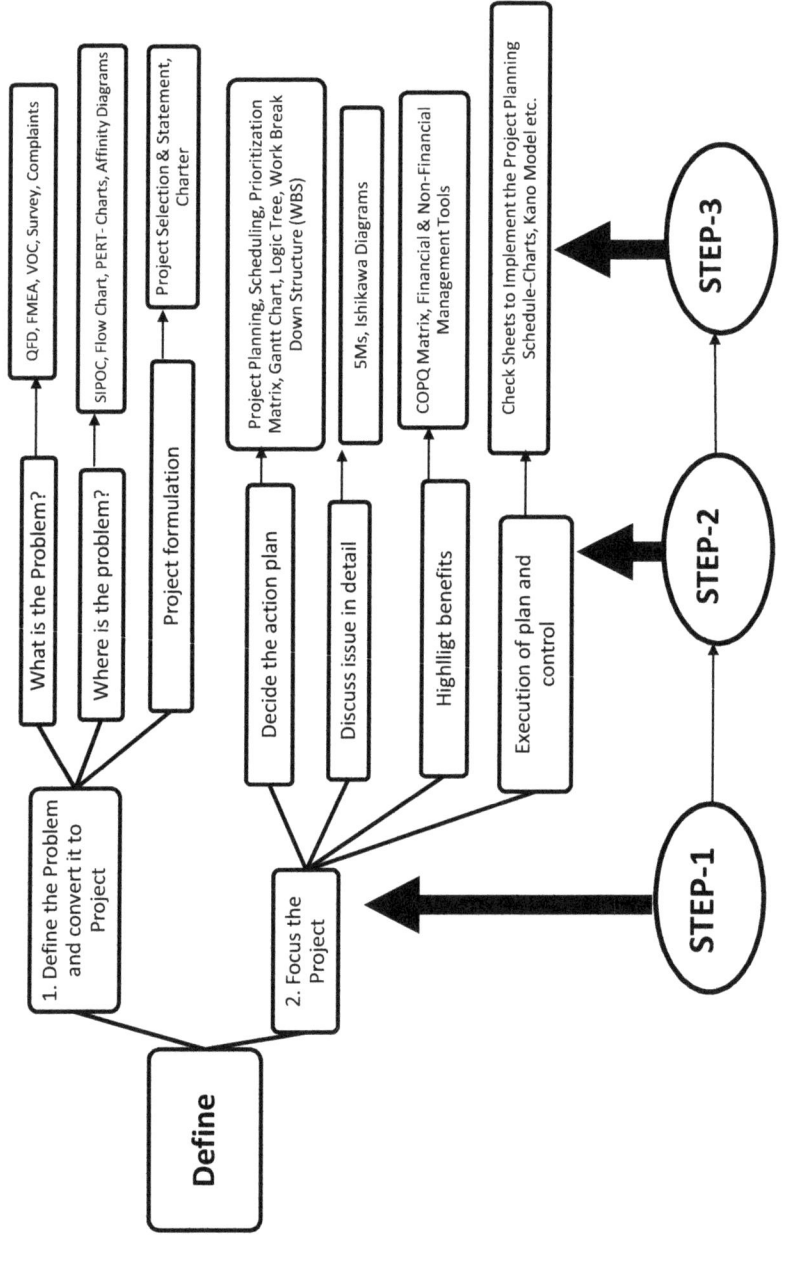

culture of an organization (Frings and Grant, 2005). Culture is not projects, Black Belts, Green Belts and teams. It is how the organization thinks about process and process improvement and this begins with the leaders. Organizations are a reflection of their leadership. If the organization wants to be a Six Sigma-driven operation, the leadership needs to be visibly involved. This means leaders must be responsible for defining projects and not simply identifying an area and a theme and handing it off to the Black Belt. Leaders need to take charge of identifying the projects, determining the business case, defining the scope, establishing the goals and estimating the project timeline. This also helps to ensure that the projects selected are truly projects that are strategically important to the business.

4.4.2 Case Findings (D-Phase)

Tools Used: *QFD, Project Charter, Project Scheduling, COPQ Matrix and SIPOC Diagram (refer table 3.1)*

This phase has been implemented step by step according to proposed framework. For this the aim has been divided into two sub-aims; problem definition and focus on problem, respectively. Under problem definition, when VOC/QFD was applied then the outcome highlighted 'high scrap' as a main problem. Pistons are machined after casting in foundry, so the machine shop in the same plant becomes a customer for foundry shop. A house of quality is generated to take the voice of customer, as shown in figure 4.4. Customer expectations (Y's) were listed vertically and corresponding ways (X's) to cope up with these requirements were shown in horizontal direction. Customer requirements/expectations were further given weightage on 1 to 5 scale. Similarly target direction had also been provided to each way. Relationship matrix was defined to create the relationship rating of each factor. Overall score was summed up in respective rows and columns. From the VOC diagram it became obvious that meeting dead line schedules, timely delivery of piston castings and low casting scrap were the main three voices of the machine shop (customer) as these factors had corresponding high overall score. These could be tackled efficiently by taking measures like; reducing the scrap in foundry, technology up gradation or innovation and through efficient process control because these remedies or ways had reflected relative more score on QFD matrix. For a given foundry system, if technology remains intact then scrap reduction and process control become the sole requirements of present customer and must be taken care of.

According to frame work next question is; *where is the high scrap?* To answer this Supplier Process Input & Output Customer (SIPOC) diagram was drawn for foundry shop. The whole manufacturing process of piston casting was re-defined and key input and output

Figure-4.4 House of Quality

TARGET DIRECTION		⊖	↑	↓	↓	⊖	⊖	↑	↑	↑	⊖	
VOC (Machine Shop)	Importance to Customer	Evaultion of Vendors	Good Process Controls	Reduction in Scrap due to Casting Defects	Less Die/Machine Set up time	Implementation of Maintenance Schedules	Ensure Adherence of Quality System	Good Inspection Plans	Technology Upgradation/Innovation	Training Plans for Work Force	Efficient Scheduling to complete Production Orders	Completeness Criteria
1	Meet Deadlines/Schedules	5	M 15	M 15	H 45	H 45	M 15	H 45	M 15	M 15	L 5	M 15	230
2	Reduction in Production Cost	5	L 5	M 15	H 45	M 15	M 15	M 15	M 15	M 15	L 5	L 5	150
3	Satisfy Quality Initiatives (Overall Quality of the Product)	4	M 12	M 12	M 12	N 0	L 4	H 36	H 36	L 4	M 12	N 0	128
4	Cycle Time Reduction	5	N 0	M 15	L 5	H 45	L 5	L 5	L 5	H 45	M 15	M 15	155
5	Reduce Rework	5	L 5	M 15	H 45	N 0	N 0	M 15	M 15	M 15	M 15	N 0	125
6	Strong Information System	3	L 3	L 3	N 0	N 0	N 0	L 3	N 0	H 27	M 9	L 3	48
7	Accountabilty of Supplied Product	3	H 27	M 9	H 27	N 0	L 3	M 12	M 12	M 12	M 12	L 3	117
8	Less Dimensional Problems	5	M 15	H 45	L 5	N 0	H 15	M 15	M 15	M 15	L 5	N 0	130
9	Develop Closer Supplier Relations	3	M 9	M 9	H 27	N 0	L 3	M 9	L 3	M 9	N 0	M 9	78
10	Reduce Waste	5	M 15	M 15	H 45	L 5	M 15	M 15	M 15	M 15	H 45	H 45	230
	IMPORTANCE RATING OF X's		106	153	256	110	75	170	131	172	123	95	
	TARGET VALUE OF CTC (Critical to Customer) FACTOR				10%								

Symbol	Relationship Between X & Y	Rating
H	Strong (H)	9
M	Medium (M)	3
L	Weak (W)	1
N	No Relation (N)	0

Targer Directions	
More is Better	▲
Less is Better	▼
Specific Amount	⊖

variables were chalked out at each production step. Full SIPOC diagram showed the whole process in terms of flow chart and defined the criticality of each factor symbolically, as shown in figure 4.5. It also explained vital process parameters with specifications (refer Annexure 1). After analyzing SIPOC diagram, it was clear that the input and output key factors for casting operation were influencing the casting scrap more than any other operation. So casting operation remained in focus and had been found as the weakest point due to which major scrap was being produced. This operation had been further divided into related sub operations to define this bottleneck completely, because one should know each and every aspect of process before trying to control it. Department of store and external vendors acted as suppliers as they supplied raw material and consumables during working.

Figure-4.5 SIPOC Diagram for Casting Process

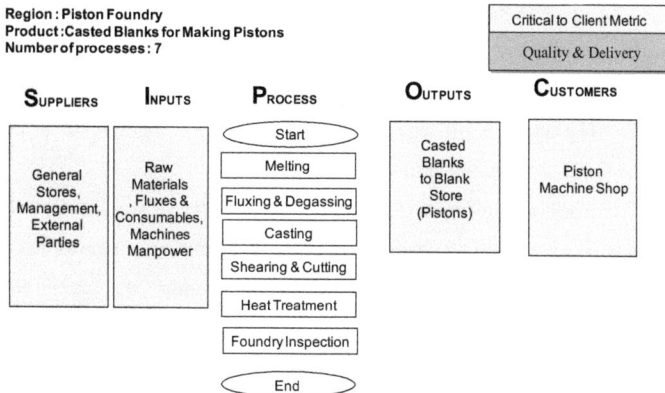

After defining the problem comprehensively, question of converting it into feasible project was raised. This is well explained by project charter as described in figure 4.6.

Figure-4.6 Project Charter

Six Sigma Project Charter					
General Information					
Project Name: Reduction in Scrap due to Foundry Defects				Project No: 5A	
				Project Location: Patiala	
Business Case:					
The piston foundry is producing 1 Lac Piston every month for North american Aftermarket. Each Scrap Piston accounts for approx Rs 183. Material Scrap in Machine Shop in Part numbers H699,H749 & H273 is very high.					
Project Details					
Problem Statement:					
From Last 1 year Foundry Rejection in Machine Shop for Part numbers H699,H749 & H273 is 20% resulting in Less Productivity ,Excess Labour Cost and Material loss accounting for Approx.35 lacs per Annum.					
Goal/Objective:					
To reduce rejection in Part numbers H699,H749 & H273 from 20% to 10%.					
Project KPOV(s): Scrap					
Project Scope/Constraints:					
Piston Foundry, Piston Machine Shop (Export) & Final Inspection will run tests only for one die Set H-749, if reduction is achieved substentiall then it will be transferred for all the H-family pistons.					
Resources					
Sr. No.	Name	BB/GB/Member	Role		Initials
1	Bikaram Jit Singh	Researcher	Six sigma Research Trainee		BJS
2	Daljeet Singh	BB	Production Supervisor		DS
3	Ravinder Sood	GB	Process Owner		RS
4	Devinder Singh	GB	SQC Supervisor		DS
5	S.K.Saini	BB	Die Repair Section		SKS
6	Harvinder Singh	Member	Technician		HS
Benefits					
Expected Bottom line Savings (Rs. lacs)			30.1 lakhs		
Notional Savings (Rs. lacs)				**Start**	
				Define	01/07/10
Strategic benefits:	Financial savings & increase in productivity			Measure	16/7/10
				Analyse	16/8/10
Other benefits:	Energy savings			Improve	08/10/10
				Control	23/11/10
				Closure and handover	01/01/11
Authorisation					
Champion			Finance Representative		
BB	Sign		Process Owner	Sign	
	Name:			Name:	

Once the charter had converted the given problem into project, then the other sub-aim was to focus on the selected project. First of all, for systematic execution of the project, proper action plan was chalked out by conducting brainstorming sessions. After dwelling on the future course of action, all possible SSVs responsible for poor casting process needed to be highlighted. The 5M concept was also carried out and it was helpful in determining the reasons for foundry scrap. Some of these findings are:

Method: The methodology/process adopted in the foundry shop primarily resulted in scrap. The various factors which resulted in the scrap as output were like turbulence during the pouring of molten metal, interruption at the time of pouring, poor molten metal treatment, poor cleaning of the ladle, furthermore ladle alignment being not proper, slow pouring speed etc. (refer figure 4.7).

Medio Ambiente: This is a Spanish word for environment. This includes climate, terrain and noise/distraction and runway environment. These external environmental forces can vary and must be considered when assessing risk of high rejection in foundry SMEs. These factors can be:

- *Climactic:* It includes temperature, seasons, precipitation, aridity and wind etc.
- *Operational:* It means routes, surfaces, terrain, vegetation, obstructions and constrictions.
- *Hygienic*: It implies vent, noise, toxicity, corrosives, dust and contaminants.
- *Vehicular/Pedestrian*: It has effect like paved, gravel, dirt, ice, mud, dust, snow, sand, hilly and curvy.

In the foundry shop it was observed that too much dust in the building and water leaks on the dies were also resulting in the scrap output.

Material: Material selection and its condition play a vital role in overall success rate of the casting process. It was observed that the major reasons of scrap could be the dirty ingots from the supplier and high temperature of the metal. Furthermore the contamination of metal with the ceramic fiber from the lids and mismatch of dies also resulted in scrap.

Manpower: It includes training, selection, proficiency, habit patterns, performance and various other personal factors of work force. The importance of manpower in manufacturing industries is quite obvious and in this context, some of the factors resulting in the scrap were;

poor inspection of piston defects, improper metal cleaning practices and keeping of furnace lids open due to negligence.

Machine: The machine category includes its design, maintenance history, performance and user perceptions. As far as machines are concerned, some reasons for scrap castings were; lack of facilitation among mating machine parts, poor die venting passages and use of dirty crucibles.

Figure-4.7 5M Diagram for Foundry Scrap

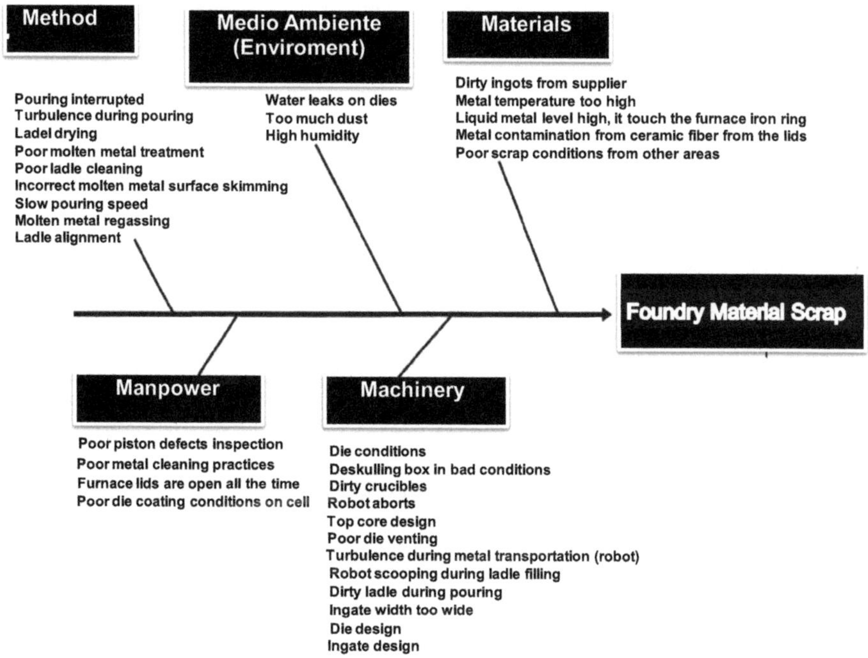

Method
Pouring interrupted
Turbulence during pouring
Ladel drying
Poor molten metal treatment
Poor ladle cleaning
Incorrect molten metal surface skimming
Slow pouring speed
Molten metal regassing
Ladle alignment

Medio Ambiente (Enviroment)
Water leaks on dies
Too much dust
High humidity

Materials
Dirty ingots from supplier
Metal temperature too high
Liquid metal level high, it touch the furnace iron ring
Metal contamination from ceramic fiber from the lids
Poor scrap conditions from other areas

Manpower
Poor piston defects inspection
Poor metal cleaning practices
Furnace lids are open all the time
Poor die coating conditions on cell

Machinery
Die conditions
Deskulling box in bad conditions
Dirty crucibles
Robot aborts
Top core design
Poor die venting
Turbulence during metal transportation (robot)
Robot scooping during ladle filling
Dirty ladle during pouring
Ingate width too wide
Die design
Ingate design

→ **Foundry Material Scrap**

The 5M tool helps to explore critical aspects or process parameters which may be influencing casting process and causing defects in castings. For this scrap and rework data can be collected from daily production reports. The tentative (or targeted) goals can be quite helpful to assess the financial benefits that would be incurred after successful implementation of DMAIC project. The COPQ matrix demonstrates net cost in terms of money because of rejection/scrap or reworking. There are three types of H pistons namely; H-273, H-519 and H-749. Out of these, H-273 had more monthly scrap as compared to other two types and hence these were selected to carry out DMAIC project for needed scrap reduction. (refer table 4.2).

Table-4.2 COPQ Matrix

Financial Parameters	H-273	H519	H749
Average Scrap/Month	1866	1100	885
Scrap Cost/Piston	183	183	183
Total Rejection Cost	341,478	201300	161955
Total Rejection Cost /Year	4097736	2415600	1943460
Total Rejection Cost for H family Pistons Annually	8456796		

COPQ matrix then calculated the net loss as Rs. 84.5 lakhs (in Indian Rupees) per annum for all H-type pistons at existing scrap levels respectively. It motivated to carry forward the selected project for feasible improvements. In the last stage of D phase, schedule chart was prepared to help in commencement of four other phases of DMAIC project. Schedule charts can be generated in terms of any sensitive parameter which has a capacity to control and motivate the whole project. In present case study, schedule chart was made in terms of days/dates as shown in table 4.3. This simple chart was a reminder on the time spent on various activities and gives a sense of achievement after completion of every single phase on-time.

Table-4.3 Schedule Chart

Phases	Scheduled Dates		Actual Dates	
	Start date	End date	Start date	End date
Define	01/07/10	15/7/10	01/07/10	15/7/10
Measure	16/7/10	15/8/10	16/7/10	15/8/10
Analyse	16/8/10	07/10/10	16/8/10	22/9/10
Improve	08/10/10	22/11/10	23/9/10	30/11/10
Control	23/11/10	31/12/10	01/12/10	31/12/10

4.4.3 Inferences from D-Phase
Some of the major inferences from this phase are:
- This phase recognized the machine shop as internal customer in present case.
- VOC matrix raised not only the voice of customer but also recommended the way to satisfy the customer through scrap reduction.
- SIPOC diagram mapped the whole foundry process in detail to understand the various possible options to achieve the prefixed goal.
- Project charter converted the customer needs into suitable project with feasible resources in hand.
- COPQ matrix predicted the expected losses of Rs. 84.5 lakhs at existing levels of scrap and also scheduled the whole project appropriately, with the help of gantt chart of six months.

4.5 Measure Phase
This phase involves an intense analysis to understand the reasons of failure of Six Sigma particularly in SMEs and it has been found that doing 'wrong operation measurement' is one of the major reasons (Singh & Khanduja, 2012e). This book proposes a more applicable methodology to execute operation measurements comprehensively, as far as their industrial applications are associated. A road map for short listing the relevant measurement tools has been suggested for six sigma learners, people working in industry and professional statisticians. The case study has validated the suggested approach successfully. Basically it is utilizing an integrated advantage of both quantitative and qualitative measurement-tools for producing breakthrough results quickly in real Indian industrial environments. Operation/process measurement is one of the vital phases, which comes into picture immediately after defining the problem (Brewer, 2004). Operation measurement though being a critical phase, is usually being neglected by most of the enterprises especially the SME sector. During literature review, various problems being faced in the measurement phase are;
- ***Lack of awareness:*** Education plays a key role in creating awareness. Education system in a country like India does not focus on the importance of measurement from an industrial point of view. Absence of any collaboration between SMEs and various

research institutes and universities, reduces awareness level significantly (Sekhar and Mahanty, 2006). This may cause wrong selection of key process metrics and further data may be collected in some non-compatible form with respect to present production environments.

- ***Lack of skilled manpower:*** Skilled workers bring degree of expertise in the performance of a given job. Workers with specialized skills in measuring instruments and formal test methods (that define the instrument's usage etc.) are in short supply and it becomes a shortfall that marks another obstacle to the global economic recovery (Hoerl, 2004). Older and experienced workers are retiring and their younger replacements often do not have the right training because their schools are out of touch with modern business needs.

- ***Non-application of high precision tools and equipments:*** With changing times and advancement in technology, measurement tools and equipments have also been upgraded (Hsu et al., 2008). In industrial metrology, highly précised measurement equipments are not being used (Snee, 2004) and due to their non-application, there are several issues on accuracy that restrict the usability of advanced metrology methods. This results in a poor manufacturing environment needing high initial investment. These factors create a major road-block in an industry's quality and quantity standards (Hoerl et al., 2001). Not only this, work force should have an appropriate awareness about the measurement system analysis (MSA) and its elements like; bias, linearity, gauge R&R and stability of measuring equipments.

- ***No calibration:*** All measuring instruments are subject to varying degrees of error and measurement uncertainty (Yang and Yeh, 2007). Hence calibration is an essential process that needs to be done after a constant interval of time, with new instrument or with sudden changes in internal/external surroundings. Though calibration techniques are expensive, but these are crucial and need to be adopted for improved and accurate results.

- ***Lack of R&D:*** Many SMEs do not have the capacity to carry out the required research on their own, either because they have little or no in-house R&D capacity or because their R&D facility lies in some another domain or sector (Nonthaleerak and Hendry, 2006). Research requires a separate budget allocation, which managements usually avoid especially in SMEs. These units lag behind in the upcoming standards and hence are not able to compete in the market.

- *No investment in re-training:* Imparting re-training to the workforce is essential for improved growth. Re-training helps the manpower to re-enhance their technical skills and knowledge about the new up gradation of measuring equipments.

Six Sigma's DMAIC methodology is a structured approach and has number of management and statistical tools/techniques to complete all its five phases comprehensively, but almost 19% of the projects have failed because of selecting wrong and absurd tools/techniques in 'Measure Phase' itself (Johnson and Swisher, 2003). So to demystify this confusion and further support the right selection criteria for various tools/techniques while implementing measure phase, an integrated methodology has been proposed and validated successfully in the case study.

4.5.1 Proposed Framework for M-Phase

The dictionary meaning of word 'Measure' is a number or quantity that records a direct observable value or performance (Antony, 2004b). The goal of this phase during Six Sigma is to get as much information as possible on the current process so as to fully understand both, how it works and how well it can work. A number of quantitative and qualitative tools/techniques are available for measuring. To help in right selection of right tool/technique, a frame work has been designed in which all the techniques/tools are further classified under four well defined steps of 'Process/Operation Measurement'. The steps are;

- Measure the process metrics
- Measure the 'As-is' operations
- Measure the SSVs
- Measure the existing measurement systems

According to the stated methodology of operation measurement, the first step is to define the process metrics, because identification of right metrics is quite helpful to collect the right data and reduce the probability of time consumption and costly errors in the path of overall operation measurement (refer figure 4.8). It is actually a unit of measurement that provides a way to objectively quantify a process. It creates a scorecard for assessing performance of Six Sigma projects and is necessary to ensure that the decisions are made on the basis of facts. Tools like Check Sheets, Tally Charts, Data Sheets, Scatter Plots, Histograms, Graphical

Figure-4.8 Proposed Framework for M-Phase

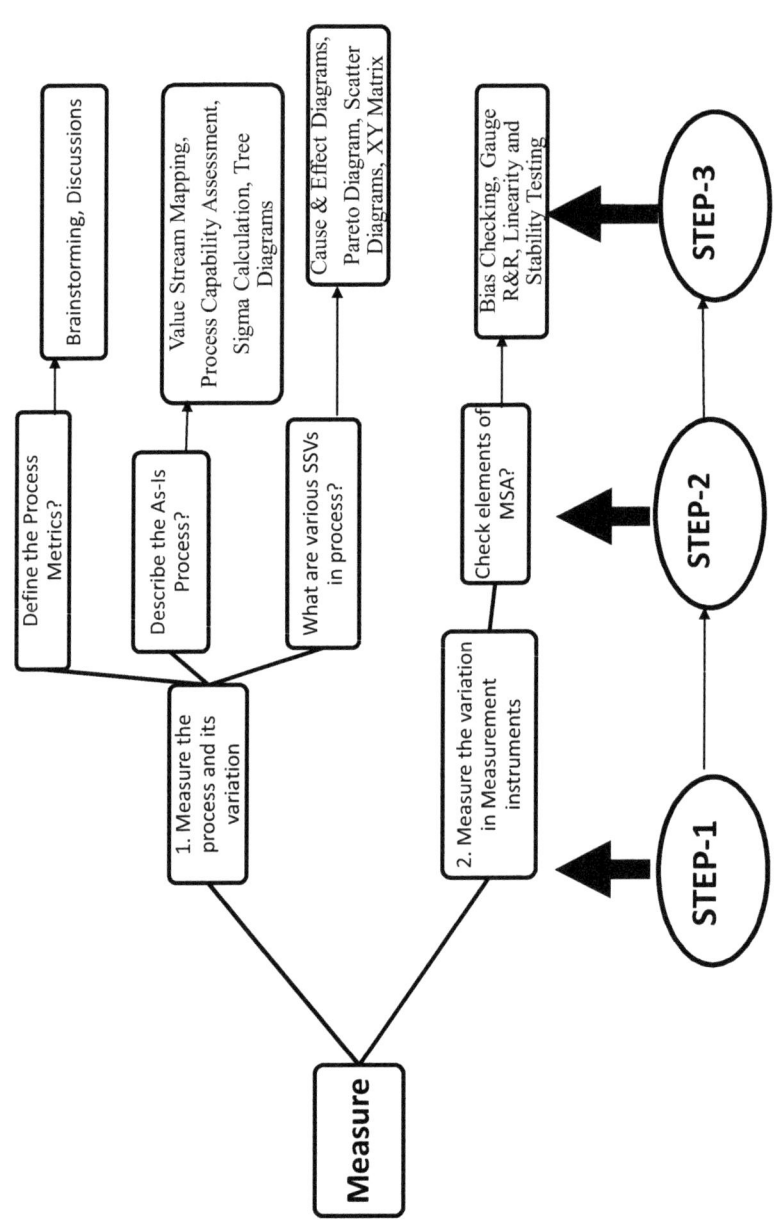

Analysis, Flow Charts etc. can be used during 'As-is' measurement of process (Jeffery, 2005). Some lean tools like Process Mapping, SIPOC Diagrams, Value Stream Mapping and Process Capability (Cpk value) can be very effective to assess the existing operations (Thomas and Barton, 2006). In third step, CTQ and SSVs within the processes should be targeted. In this step, data collection being the major goal needs to be done systematically with careful planning. The primary focus of data collection should be in gathering data that helps to describe the problem, as well as uncovering any factors that provide clues about how, when, where or in what circumstances the problem occurs or is worsened. In the last step, validity of measurement equipments needs to be checked by implementing various elements of MSA. It may contain bias checking, linearity test of measuring gauge, stability monitoring of equipment or calculation of gauge's repeatability and reproducibility (Gauge R&R). At the end of this phase, the project team should have detailed process map, broad baseline data and Sigma value of the given business process. Team members should have a clear understanding of the existing process and its various CTQ issues (Antony et al., 2007).

4.5.2 Case Findings (M-Phase)

Tools Used: *Cpk Assessment, Sigma Calculator, Pareto Chart, 5Ms, VSM, C&E Matrix, Bias checking, Gage R&R and Stability Test (refer figure 3.1)*

As per the proposed methodology, during the first step it was realized that the 'percentage of scrap' is a useful metric to monitor the quality of operations. Pistons being produced were of three types; H-273, H-519 and H-749 (as discussed earlier) and due to high scrap rate of around 22%, their casting process was observed and has been mapped in figure 4.9. The basic aim of the value stream mapping is to minutely measure the bottleneck process. Each operation is drawn with standard activity symbols and prescribed process number with its description. Key process input variables at each operation have been quoted and further classified in three categories; Noise, Critic and Controlled (non-critic). Noises are the operation parameters that can't be controlled due to some specific reasons or for whom there is no need to control because they don't have any significant effect on the process metric. Non critics are the process parameters which can affect the metrics positively, but they are already in well control. The entire critic or critical to quality (CTQ) process parameters have also been highlighted during process mapping of foundry shop because these are actually under focus and can give desired breakthroughs. After analyzing the monthly scrap reports of these pistons, it was deduced that the major critical dimension in pistons is 'bottom thickness' (refer figure 4.10) and its variation should be maintained in the tolerance limits of 9.7±0.25mm.

Figure-4.9 Process Mapping of Casting Process

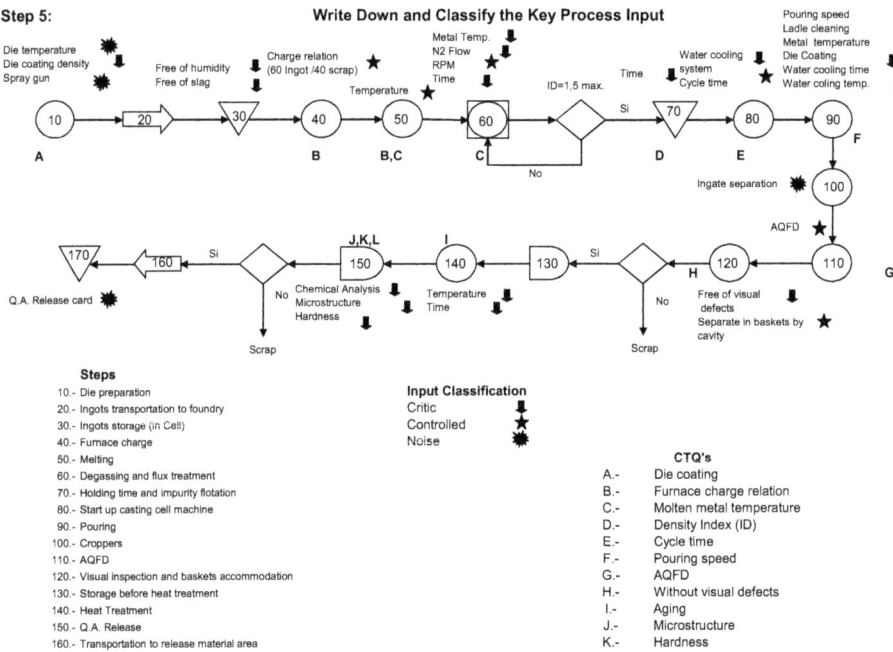

Figure-4.10 Pictorial Presentation of Bottom Thickness Variation

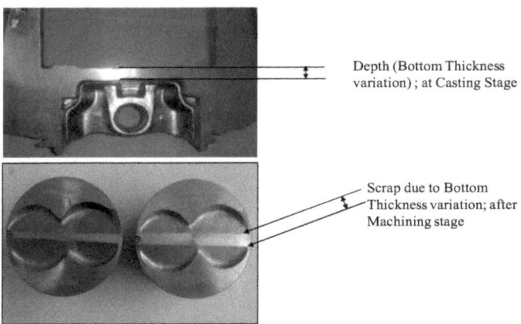

At the end of 'As-is' step, process capability (Cpk) of casting process was calculated in terms of bottom thickness variation, as shown in figure 4.11. It was coming out to be 0.21. It is very low, as large area of normal curve is falling outside the upper and lower limit of curve and hence indicating that the existing casting process had very less control over the bottom thickness dimension. The lower and upper specification limits were 9.54 and 9.95 respectively.

Figure-4.11 Calculation of Cpk Value in Terms of BT Variation

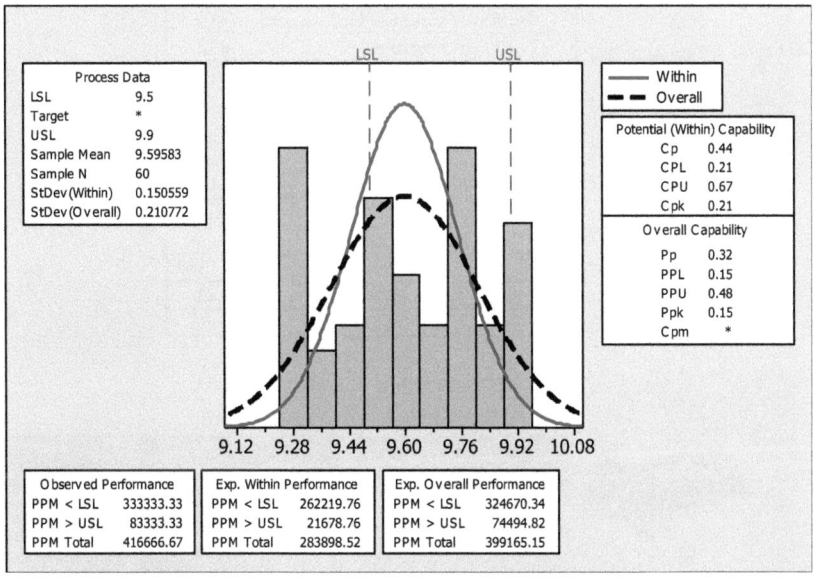

The mean has been generated as 9.59 with standard daviation 0.210. The other process indices like Cp, CpL and CpU were 0.44, 0.21 and 0.67 respectively. Minitab has also calculated parts per million capabiliy (Ppk) as 0.15 which is very less (see annexure 2 for data details). This Cpk study has only taken 'scrap due to BT variation' into account while calculating the casting process capability. Due to this proper sigma level of process could not be judged accurately, as there existed six other reasons for scrap. To measure the overall sigma value of casting process a 'Six Sigma Calculator' (a soft programme, based upon defect per million opportunity) was used. Total monthly production and different types of scrap were uploaded into sigma calculator (refer figure 4.12). The scrap data of June-2010 for pistons had been fed. The defect per million opportunity (DPMO) was calculated as 26770 and corresponding sigma value of casting was comit out to be 3.4. It is very low and alarming for an ISO-16479 foundry. Major casting defect in H-273, H-749 and H-519 pistons have been shown in figure 4.13.

Figure-4.12 Calculation of Sigma Value with Sigma-Calculator

Production results of June-2010	
Total number of Machined parts	8482
Scrap type	**Nos.**
Bottom Thickness Defect	480
Blow Holes	432
Cold Lap	120
Depression	120
Hydrogen Porosity	48
Shrinkage	457
Defective Pin Hole	210
Total Scrap in June	1866
Nos. of Opportunities	7
DPMO	31428
Sigma Level of Process	**3.43**
Yield (%)	76.96

Figure-4.13 Snaps of Casting Defects

Figure-4.14 Pareto Chart

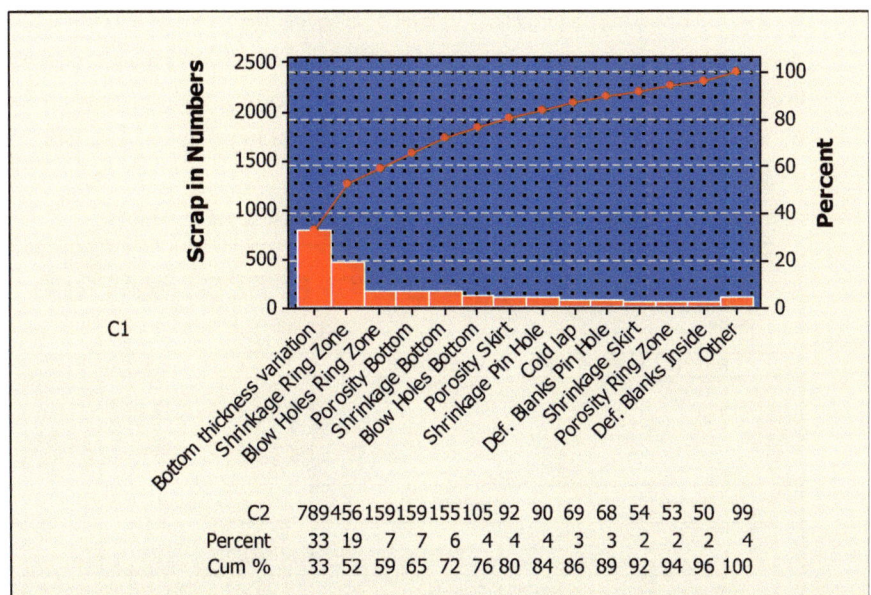

Pareto chart for H-family pistons was drawn to give statistical picture of casting scrap in context with corresponding reasons/causes (see figure 4.14). It is a type of chart that contains both bars and a line graph, where individual values are represented in descending order by bars and the cumulative total is represented by the line. The purpose of this chart is to highlight the most important among a (typically large) set of factors. Pareto charts have measured shrinkage, porosity, blow holes, pin holes and bottom thickness as the main reasons of scrap. After measuring the present operations, SSVs for poor casting process needed to be highlighted. The root cause-analysis was performed to explore for scrap where each cause or reason for imperfection is a source of variation. Every factor/reason of scrap has been given weight ages on 10 point scale (refer figure 4.15). The relation of each reason with respective root cause has been shown by allocating a number on 5 point scale in concerned cell and then total sum of numbers for each root cause had been calculated. Depending upon the high value of total score, all the root causes (parameters) were categorized as Critic, Control and Noise. The matrix makes it clear that process parameters like gate feeding design, dimensional accuracy of die casting machine, die temperature and preheating, die coating thickness, discharge of cooling water, non-continuity in casting process, metal sticking on pins and skill level of operators were some of the critical SSVs in casting process and were responsible for around 20% of scrap due to various reasons/factors like bottom thickness variation, shrinkage, blow holes, porosity, cold lap and pin holes etc. By now eleven CTQ factors were shortlisted that seemed to be the major reasons of high scrap.

Figure-4.15 Root Cause Analysis by Cause & Effect Matrix

Characteristic										
Critic		**CAUSE AND EFFECT MATRIX**								
Control										
Noise										

	RATING OF IMPORTANCE TO CUSTOMER		10	9	9	6	6	10	9		
	EFFECTS		Defective blanks bottom	Blow Holes	Cold Laps	Depression	Hydrogen Porosity	Shrinkage	Defective Pin Hole	Total	Actions Decided
S.NO	Process Step	Process Input	Correlation of Input to Output								
1	Design	Bottom Seating Design	5	0	0	3	0	0	0	68	
2	Design	Gate Feeding Design	3	3	3	0	2	5	0	146	
3	Die repair	Dimensional accuracy of Die parts	5	0	2	5	0	0	0	98	
4	Die repair	Dimensional accuracy of casting Machine parts	4	0	1	3	0	0	0	199	
5	Set up	Die Temperature & Preheating	0	5	4	0	5	5	2	179	
6	Set Up	Die Cooling Connections	0	5	4	0	4	4	0	145	
7	Set Up	Die Coating Thickness	4	4	4	0	4	4	0	176	
8	Set up	Water Cooling Pressure too low	0	3	0	0	5	4	0	97	
9	Set up	Discharge of Cooling water inside die parts	0	5	3	0	3	5	0	140	
10	Set up	Vaccum Pressure in air vents	0	3	5	0	4	2	0	116	
11	Set up	Ratio of water to Dycote	0	3	4	0	3	2	0	101	
12	Casting	Low Pouring Speed	0	2	5	0	0	3	0	93	
13	Casting	Die Core grouping	3	0	0	2	0	0	0	42	
14	Casting	Non-continuity in casting process	0	3	5	2	5	5	0	164	
15	Casting	Metal Sticking on Pin	0	0	0	5	0	0	5	75	
16	Pouring	Casting Temperature too High	0	4	0	0	5	5	0	116	
17	Pouring	Casting Temperature too Low	0	4	3	0	5	5	2	161	
18	Pouring	Degassing Procedure & waiting time	0	5	2	0	4	5	0	137	
19	Pouring	Skill of Operator	3	1	3	1	2	2	2	122	
			270	450	432	126	306	560	99		

The next crucial step was the measurement of accuracy and precision of already 'in-use' measuring equipments or gauges. It was decided to validate the calibration by conducting Gage R&R study for bottom thickness gauge (BT gauge), as its precision started deteriorating after a span of one month itself and scrap due to variation in bottom thickness of piston was also very high. The data collected for gauge R&R of BT gauge has been channelized (refer table 4.4) and ten piston parts in random were selected and their bottom thickness was checked with BT gauge by three operators repeatedly.

Table-4.4 Random Data Collection Plan for Conducting Gage R&R

PART NAME	PISTON			GAGE NAME:		BT CHECKING INSTRUMENT						
CHARACTERIST BOTTOM THICKNESS				GAGE NO.:		MBT-03 L.C 0.01 MM						
SPECIFICATION: 0.40 MM												
					DATA SHEET							
					PART							AVERAGE
APPRAISER	TRIALS	1	2	3	4	5	6	7	8	9	10	
AVTAR SINGH	1	8.12	7.88	8.16	8.00	7.81	8.12	7.88	7.87	8.13	8.10	
	2	8.10	7.86	8.16	8.00	7.80	8.11	7.87	7.88	8.12	8.10	
	3	8.11	7.86	8.13	8.02	7.81	8.10	7.87	7.88	8.12	8.10	
	Average	8.11	7.87	8.15	8.01	7.81	8.11	7.87	7.88	8.12	8.10	X_a= 8.00
	Range	0.02	0.02	0.03	0.02	0.01	0.02	0.01	0.01	0.01	0.00	R_a= 0.01
NACHATTAR SINGH	1	8.12	7.88	8.16	8.02	7.81	8.12	7.88	7.87	8.13	8.10	
	2	8.10	7.87	8.14	8.00	7.80	8.11	7.86	7.86	8.12	8.11	
	3	8.11	7.86	8.13	8.02	7.80	8.10	7.87	7.88	8.11	8.10	
	Average	8.11	7.87	8.14	8.01	7.80	8.11	7.87	7.87	8.12	8.10	Xb= 8.00
	Range	0.02	0.02	0.03	0.02	0.01	0.02	0.02	0.02	0.02	0.01	R_b= 0.02
MOHINDER SINGH	1	8.10	7.88	8.16	8.00	7.81	8.12	7.88	7.87	8.13	8.10	
	2	8.10	7.90	8.16	8.00	7.82	8.11	7.90	7.88	8.12	8.11	
	3	8.12	7.86	8.13	8.02	7.81	8.12	7.87	7.88	8.12	8.10	
	Average	8.11	7.88	8.15	8.01	7.81	8.12	7.88	7.88	8.12	8.10	X_c= 8.01
	Range	0.02	0.04	0.03	0.02	0.01	0.01	0.03	0.00	0.01	0.00	R_c= 0.02
16. Part Ave X_p		8.11	7.87	8.15	8.01	7.81	8.11	7.88	7.87	8.12	8.10	Rp= 0.34
17 Ra+Rb+Rc/N of Apraiser												R= 0.0170
18 Max X-Min X=X_{diff}												Xdiff= 0.0047
19 R* D_4 =UCL_R												UCL 0.044
20 R* D_3 =LCL_R												LCL 0

At 95% confidence level, ANOVA of two-way interaction or without interaction has shown 'p' values above 0.05 and hence it confirmed the correctness of the operator and part to part variation of BT gauge. The repeatability came out to be 48 (more than 30) and hence it was well in control (refer figure 4.16).

Figure-4.16 Gage R&R Measurement

Gage name:	B.T. Gauge
Date of study:	02/08/10
Reported by:	Bikram Jit singh
Tolerance:	0.40mm

```
Two-Way ANOVA Table With Interaction

Source              DF      SS          MS          F         P
PartID               9    0.948327    0.105370   1433.24    0.000
Operator             2    0.000243    0.000122      1.65    0.219
PartID * Operator   18    0.001323    0.000074      0.92    0.564
Repeatability       30    0.002400    0.000080
Total               59    0.952293

Alpha to remove interaction term = 0.25
```

Two-Way ANOVA Table Without Interaction

```
Source            DF      SS          MS          F         P
PartID             9    0.948327    0.105370   1358.39    0.000
Operator           2    0.000243    0.000122      1.57    0.219
Repeatability     48    0.003723    0.000078
Total             59    0.952293
```

Gage R&R

```
                                    %Contribution
Source              VarComp         (of VarComp)
Total Gage R&R      0.0000798           0.45
  Repeatability     0.0000776           0.44
  Reproducibility   0.0000022           0.01
    Operator        0.0000022           0.01
Part-To-Part        0.0175487          99.55
Total Variation     0.0176285         100.00

                                    Study Var    %Study Var
Source              StdDev (SD)     (6 * SD)      (%SV)
Total Gage R&R       0.008932        0.053590      6.73
  Repeatability      0.008807        0.052844      6.63
  Reproducibility    0.001485        0.008909      1.12
    Operator         0.001485        0.008909      1.12
Part-To-Part         0.132471        0.794829     99.77
Total Variation      0.132772        0.796633    100.00
```

Number of Distinct Categories = 20

Figure 4.17 explains the results of gage R&R graphically. The first graph is regarding component variation as the repeatability and reproducibility of gauge is coming into picture. The second one is a measure of bottom thickness variation among each piston parts. Third and fifth is a representation of range and mean plot of BT gauge output. Fourth graph explains the operator to operator variation for same piston part measurements. Last graph shows collective variation of bottom thickness due to operator and part ID, together. The overall results seem to be positive and indicate that the BT gauge has enough repeatability and reproducibility in the present case and needs no calibration.

Figure-4.17 Graphical Representation of Gage R&R Results

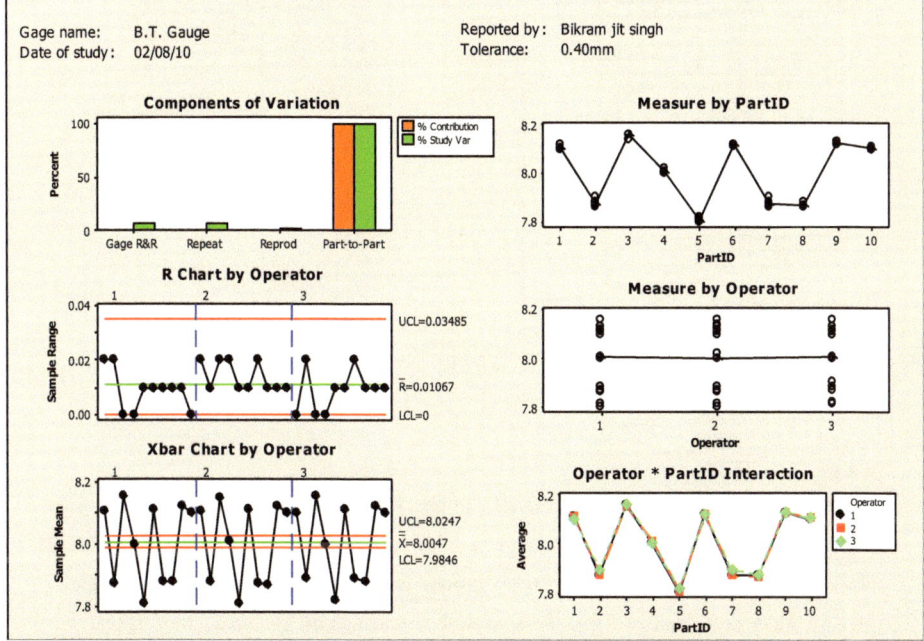

During the last step of proposed methodology, bias test of immersion pyrometer was conducted, because molten temperature can be the main reason of different casting defects. In this study difference between observed average of measurements and reference value was taken. Reference value, also known as accepted reference value, is a value that serves as an agreed upon value of measured values. This reference value can be determined by averaging several measurements with more precise measuring equipment. Fifteen trials were conducted by measuring temperature of molten metal in a holding furnace. The difference of each trial with the master value (or Bias) was calculated and plotted, Because both upper and lower limits were coming on negative side only (or zero value is not lying in between), so pyrometer seemed to be biased on lower side of temperature readings and so needed immediate repair or setting (refer figure 4.18).

Figure-4.18 Bias Study of Immersion Pyrometer

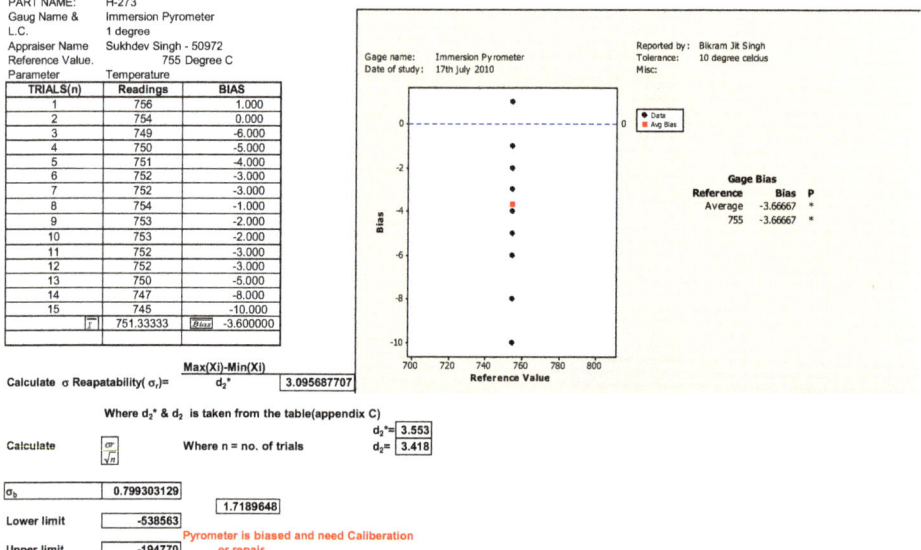

PART NAME:	H-273
Gauge Name & L.C.	Immersion Pyrometer 1 degree
Appraiser Name	Sukhdev Singh - 50972
Reference Value.	755 Degree C
Parameter	Temperature

TRIALS(n)	Readings	BIAS
1	756	1.000
2	754	0.000
3	749	-6.000
4	750	-5.000
5	751	-4.000
6	752	-3.000
7	752	-3.000
8	754	-1.000
9	753	-2.000
10	753	-2.000
11	752	-3.000
12	752	-3.000
13	750	-5.000
14	747	-8.000
15	745	-10.000
$\overline{T}$	751.33333	$\overline{Bias}$ -3.600000

Calculate σ Repeatability (σ_r) = $\dfrac{Max(Xi)-Min(Xi)}{d_2^*}$ = 3.095687707

Where d_2^* & d_2 is taken from the table (appendix C)

Calculate $\dfrac{\sigma_r}{\sqrt{n}}$ Where n = no. of trials

d_2^* = 3.553
d_2 = 3.418

σ_b = 0.799303129

1.7189648

Lower limit = -538563
Upper limit = -194770

Pyrometer is biased and need Calibration or repair

Lastly stability of Vac-tester was evaluated to check the working of this significant tester over a long period of time (refer table 4.5). This tester ensures proper de-gasification of molten metal before starting the die casting process and faulty tester may cause scrap due to porosity, blow holes and cold lap. Randomly four readings of Vac-tester were taken after every three hours during three shifts and arranged in tabular form. By using excel sheet, stability of the tester was calculated as 28.26% only and it was not acceptable. If upper and lower limit values are coming positive or on one side of zero, it implies stability is not symmetric about zero value and needs calibration. The graphical analysis of Vac-test readings, for all the twenty days, was plotted on suitable scale and control limits. The graphs shows tester was stable from first day of month to 11th day but due to some special reasons it became un-stable and went beyond upper control limit up to the first half of 12th day. It implies that for at least four shifts, tester was running in unstable conditions and was responsible for scrap to some extent.

Table-4.5 Data Collection and Measurements for Stability Test

Date	Feb	1	3	4	5	6	7	8	10	11	12	13	14	15	18	19	20	21	22	24	25
Time	X1 N/s																				→
	X2 M/s																				
	X3 A/s N/s																				
	X4																				
		1	2	3	4	5	6	7	8	9	10	11	12	13	14	15	16	17	18	19	20
Readings	X1	0.4	0.3	0.4	0.6	0.3	0.4	0.4	0.4	0.5	0.4	0.3	0.3	0.4	0.3	0.4	0.1	0.6	0.4	0.4	0.3
	X2	1.2	0.2	0.3	0.4	0.4	0	0.5	0.3	0.3	0.3	0.4	1.2	0	0.4	0.3	0.2	0.4	0.5	0.4	0.4
	X3	0.5	0.3	0.4	0.4	0.4	0.2	0.3	0.5	0.4	0.3	0.5	0.2	0.4	0.3	0.5	0.2	0.4	0.4	0.3	0.3
	X4	0.2	0.4	0.3	0.4	0.4	0.4	0.4	0.4	0.4	0.6	0.5	2.3	0.5	0.4	0.5	0	0.5	0.4	0.4	0.3
Xbar		0.5750	0.3000	0.3500	0.4500	0.3750	0.2500	0.4000	0.4000	0.4000	0.4000	0.4250	1.0000	0.3250	0.3500	0.4250	0.1250	0.4750	0.4250	0.3750	0.3250
Range		1.000	0.200	0.100	0.200	0.100	0.400	0.200	0.200	0.200	0.300	0.200	2.100	0.500	0.100	0.200	0.200	0.200	0.100	0.100	0.100

Sub Group Size= 4   A2= 0.729   D3= 0   D4= 2.28   D2= 2.059
Standard deviation (s) $\bar{R}$ = 0.197873597
Gauge stability=   s/Tolerance   X100   28.26765674 %

BIAS CONTROL CHART METHOD   BIAS   0.40750
ULX   =   0.6517   d2=   2.05875 $\sigma_r = \dfrac{\bar{R}}{d_2^*}$ = 0.16198208 $\sigma_b = \dfrac{\sigma_r}{\sqrt{g}}$ =   0.036220294
Xdbar =   0.4075
LCLx =   0.163   d2*= 2.06813
UCLr =   0.76   tv,1-a/2 =   2.000   Lower Limit =   0.335388
R bar =   0.3350
LCL   =   0   alpha =   0.05   Upper Limit =   0.479612034

Result   Gage is not well stable

Similarly on 16th day, tester again went beyond the lower limit for some hours and came again within limits as graphically represented in figure 4.19.

Figure 4.19 Mean-Range Control Chart for Stability of Vac-Tester

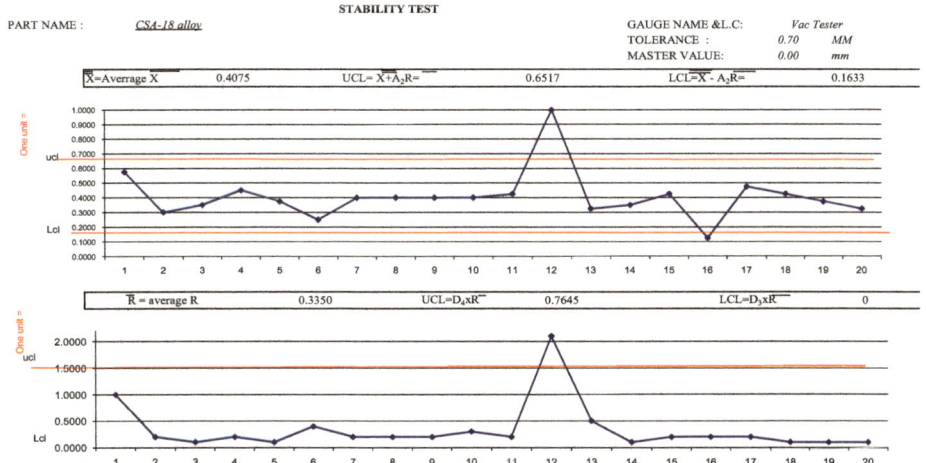

4.5.3 Inferences from M-Phase

Whole operation/process measurement activity was accomplished successfully without any misconceptions by suitable utilization of proposed methodology. Some of the major inferences from this phase are:

- Scrap had been brain stormed as the main process metric which must be focused for breakthroughs.
- Casting process mapping helped to recognize critic, non critic and noise CTQs.
- Sigma level of running process came out to be 3.37, calculated on the basis of DPMO. Captured pictures of various casting defects further cleared the reasons of scrap.
- Pareto chart had been generated to prioritize the reasons of scrap, depending upon their respective degree of severity and found BT variation, shrinkage and blow holes as major defects, causing more than 80% of scrap.
- Cause and effect diagram identified eleven SSVs responsible for these defects in castings.
- An appropriate MSA had been executed and it found bias in immersion pyrometer and un-stability in Vac tester, so necessary repair and calibration had been performed before starting the next phase and almost all possible CTQs were measured with their intensities. Gage R&R of BT Gauge had come within specifications and found ok.
- This phase has emerged as a discipline in itself and is motivated by its own program of development including integration of old and new measurement tools and techniques.
- Lot of time and efforts were saved by following suggested framework.

4.6 Analyse Phase

Six Sigma, unlike other business improvement approaches, has strong customer focus and contains key concepts related to strategy, organizational change, training and setting stretch objectives (Benedetto, 2003). The define phase establishes the rationale (Amer et al., 2007) and the measure phase involves studying and understanding the CTQs (Basu, 2004). Analysis deduces the useful information from the measured data and also checks the authenticity of each selected CTQs. It finalizes the CTQs on the basis of their significance and further helps to decide improvement strategy during next phase. This phase ensures the right input to improve phase for feasible results (Singh & Khanduja, 2011b). Weak analysis and further ignorance towards the application of various quantitative or statistical tools appears to be a prominent reason for faulty analysis. This book proposes a frame work which has classified the entire quantitative and qualitative analytical tools with respect to present production environments.

From literature survey, it appears that mostly the success of overall Six Sigma project depends on analysis as it is the final significant step to select and highlight the SSVs that may be taken further to improve-phase. So it becomes very important to conduct the right analysis of CTQs for achieving fruitful results from DMAIC approach. Mostly it has been seen that there are some basic reasons for error in analysis phase and these are:

- Less knowledge of applied statistical concepts
- Un awareness of analysis tools and techniques
- Wrong selection of analytical methods for analysis

In SMEs, managers mostly have less exposure to specific statistical tools and techniques, which are the backbone of any Six Sigma project. Moreover such people also have less knowledge regarding analysis tools and methodologies. Generally, academic views of analytical tools seem to be quite different from actual application. Due to this many practitioners of Six Sigma choose wrong (non-compatible) tools leading to absurd analysis. This may cause short listing of wrong CTQs and thus putting all the improvement efforts in vain.

4.6.1 Proposed Framework for A-Phase

The key ingredient for a successful Six Sigma management process is the commitment of top management. Executives must have a burning desire to transform their organizations. This means total commitment from the top of the organization to the bottom of organization (Thakkar et al., 2006). Six Sigma management focuses on driving effective and efficient performance across the total enterprise to increase the perception of the marketplace of its ability to deliver value-added processes, products and services (Sinan et al, 2010). It is intended to bifurcate analysis-tools in more simple and adaptable form for the direct use of any Six Sigma practitioner, so that these can be used easily at right place and on right time. Over the years there have been a large amount of complex discussions and arguments on analysis methodology and how an inquiry should proceed. Much of this debate has centered on the issue of qualitative versus quantitative inquiry–which might be better and which is more scientific (refer figure 4.20).

Different methodologies have become popular at different social, political, historical and cultural times in our development and it has been felt that all methodologies have their specific strengths and weaknesses (Vote and Huston, 2005), but present work emphasizes utilization of an integrated approach. Qualitative analysis has to do with deductive reasoning not necessarily the quantity and it explores attitudes, behavior and experiences through such methods. Quantitative analysis is the systematic scientific investigation of quantitative properties and phenomena and their relationships. Neither is better than the other – they are just different and both have their pros and cons respectively (Wright and Basu, 2008).

Figure-4.20 Traditional Classification of Analyse Tools

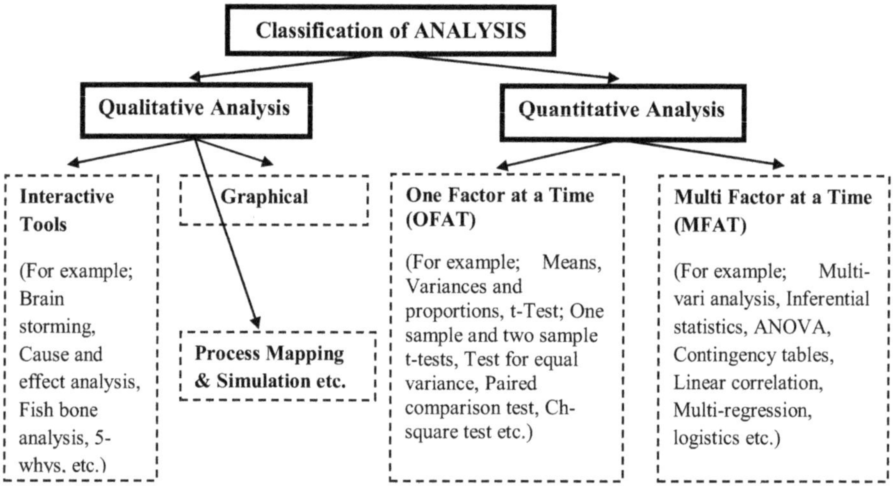

```
                    Classification of ANALYSIS
                    /                        \
          Qualitative Analysis          Quantitative Analysis
           /         \                   /              \
  Interactive    Graphical        One Factor at a Time   Multi Factor at a Time
  Tools                           (OFAT)                  (MFAT)
```

Interactive Tools	Graphical	One Factor at a Time (OFAT)	Multi Factor at a Time (MFAT)
(For example; Brain storming, Cause and effect analysis, Fish bone analysis, 5-whys, etc.)	Process Mapping & Simulation etc.	(For example; Means, Variances and proportions, t-Test; One sample and two sample t-tests, Test for equal variance, Paired comparison test, Ch-square test etc.)	(For example; Multi-vari analysis, Inferential statistics, ANOVA, Contingency tables, Linear correlation, Multi-regression, logistics etc.)

(Source; Breyfogle, 1999)

It is hard to see any existence of tool-selection-criteria for managers performing live projects in industries. For this, depending upon the type of analysis or nature of given data, a road map has been chalked out (refer table 4.6) for an appropriate selection of optimum analysis tool. This road map (criteria) will positively help to understand the nature and application of respective tools/techniques in a more practical way, which will in turn reduce the chances of poor and wrong selection of analytical tools/techniques during execution of 'Analyse Phase'.

Table-4.6 Proposed Road Map for Selection of Analytical Tools

Type of Analysis	Quantitative Tools	Qualitative Tools
Describing a group or several groups	Ordered Array, Frequency Distribution, Stem-and-Leaf Display, Relative Frequency Distribution, Percentage Distribution, Cumulative Percentage Distribution, Histogram, Polygon, Cumulative Percentage polygon.	Summary Tables, Bar-Charts, Pie-Charts, Pareto-Diagram, Matrix-Plots
Inference about one group / data population	Confidence Estimation for Means, Z-test for Means, t-Test for Means	Cumsum Plots, Process Modeling and Tree Diagrams, Marginal Plots
Comparing two groups	Test for the Difference in the Mean of Two Independent Populations, Paired t-Test, f-Test and Z-test for the Difference Between the Two Variances,	Interaction Plots and Simulation, Matrix plots, Affinity Diagrams
Comparing more than two groups	One Way Analysis of Variance (ANOVA) and Two Way ANOVA, Balanced ANOVA, Fully Nested ANOVA	Main Effect Plot, Symmetric Plot, Gage Run Charts, Bias Graphs
Analysing the relationship between two variables	Scatter Diagrams, Time Series Plots, Covariance, Coefficient of Correlation, Correlation and Simple Regression, t-test for Correlation,	Contingency Table, Side by Side Bar Charts, Probability Plots, Interval Plots
Analyzing the relationship between two or more variables	Multi-Regression, Stepwise Regression, Fitted Line-Plot, Binary Logistic Regression, Ordinal Logistic Regression, Nominal Logistic Regression, Chi-Square Test of Independence	Multi-vary Charts, Activity Network Plots, Area Graphs, Contour Plots, Line Plots
Understanding quality and process variability	Statistical applications in quality and productivity management like; Run Charts, Individual Distribution Identification, Capability Analysis (Normal, Non-Normal, Poisson and Binomial and Multi-Variable), MSA (Gauge R & R: Crossed & Nested, Bias Checking and Analysis of Stability)	Attributre Agreement Analysis, Acceptance Sampling by Attributes, Reliability Study
Understanding and solving decision problems	Decision Analysis, Design of Experiments (Full Factorial, Fractional Factorial and Tanguchi's Method), Response Surface Methodology (RSM)	Trend Analysis, Box Plot, Root Diagrams, 3-D Surface Plots, 3-D Scatter Plots

| Predictions for business planning | Statistical Forecasting Techniques | Pie & Bar Charts, Histogram, Frequency Polygons |

For example; generally frequency distribution or percentage distribution are being used as quantitative tools to analyse a group of data or number of such groups but these may be handled with cumulative percentage polygon and ordered arrays in more efficient ways. Besides this, the framework also provides some corresponding qualitative tools like; bar charts, pareto diagrams, matrix plots and summary tables to further deepen the analysis. Similarly, use of one way ANOVA and two ways ANOVA is quite common to compare more than two group populations, but this road map reminds us to see the situation and type of data in hand. It suggests options of balanced ANOVA and full nested ANOVA also to think about. Not only this, it also offers simple qualitative tools like; effect plots, bias graphs or symmetric plots to analyse the problem briskly.

4.6.2 Case Findings (A-Phase)

Tools Used: *Chi Square Test, One Way ANOVA, 2-Sample t-Test, Multi Regression, Interaction Plot, Ishikawa Diagram and Why-Why Analysis (refer table 3.1)*

At the end of Measure Phase, eleven CTQs were shortlisted, which seemed to be the major reasons of high scrap. Now before attacking these SSVs through Improve Phase, the authenticity and impact realization of each SSV on scrap was judged by conducting suitable investigation. This phase tries to further focus improvement effort on those SSVs which can lead to scrap reduction. As per proposed framework, analysis of each critical factor is done to:

- Remove the SSVs that have negligible/less impact or that may be in well controlled conditions in the present case.
- Prioritize the SSVs according to their influence on dependent variable (scrap), which in terns helps to design the action plan for improvement phase.

Due to high significance of qualitative as well as quantitative analytical techniques, a combination of both has been used for performing analysis (refer table 4.7). First step was to study the various highlighted SSVs with respect to their nature, for deciding relevant analytical feed required to tackle each critical factor. An extensive analysis was performed on the eleven identified SSVs by using various tools like; Chi square test had been decided to apply on shift-scrap data for analyzing any dependency of scrap on some specific shift.

Similarly, ANOVA had been used to found impact of variation in die coating thickness on over all casting scrap and so on.

Tables 4.7 Plan for Analyse Phase

Type of Analysis Technique	Name of Tool Used	SSVs Analysed
Hypothesis Testing (OFAT)	Chi Square Test	Shift Dependency
	One Way ANOVA	Die Coating Thickness
	2 Sample t-Test	In Gate Design
		Discharge of Cooling Water
		Operator Skill
MFAT	Multi Regression	Alloy Temperature
		Die Temperature
		Delay Time
Graphical Analysis	Interaction Plot	Vaccume Pressure
Qualitative Technique	Ishikawa Diagram	Metal Sticking on Pins
	Why-Why Analysis	Machine's Dimensional In-accuracy

These SSVs have been put in seven categories, as per the nature of tool used. These are:

Category-1

SSVs Selected: Shift Dependency

Tools Used: Chi-square Test

When there are two or more groups and in each group there are number of sub groups, then Chi-square test is used to check the variation in the mean of groups (Velazquez et al., 2010). The null hypothesis is formulated as of equal mean and alternate as of unequal one (refer table 4.8)

Table-4.8 Data of Chi-Square Test

Null Hypothesis   Ho: $\mu1 = \mu2 = \mu3$

Actual Hypothesis   Ha: $\mu1 \neq \mu2 \neq \mu3$

Shifts	Production	Good	Scrap	BT Vary	Blow Holes	Porosity	Shrinkage	Depression	Cold Lap	Pin Hole
N/S	128	102	26	13	4	1	3	2	1	2
M/S	124	100	24	8	6	2	3	1	2	2
A/S	127	101	27	10	2	3	5	2	3	2

There were three shifts (groups) and each had respective independent quantity of scrap. Next shift- scrap was subdivided into six different sub groups (i.e. scrap due to blow holes, shrinkage, pin holes etc.) Minitab has been used to calculate expected values and chi square values for respective observed values (refer figure 4.21).

Figure-4.21 Calculation for Chi-Square Values and Testing

Session Confirmation from Minitab

Expected value

(Row obs. total) * (Column obs. total)

Grand observation Total

	B.T.vary	B. hole	Prsty	Shrkg	Dep.	C. lap	P.H. defect	Total
1	13	4	1	3	2	1	2	26
	10.47	4.05	2.03	3.71	1.69	2.03	2.03	
	0.613	0.001	0.520	0.137	0.058	0.520	0.000	
2	8	6	2	3	1	2	2	24
	9.66	3.74	1.87	3.43	1.56	1.87	1.87	
	0.286	1.365	0.009	0.054	0.200	0.009	0.009	
3	10	2	3	5	2	3	2	27
	10.87	4.21	2.10	3.86	1.75	2.10	2.10	
	0.070	1.158	0.382	0.339	0.035	0.382	0.005	
Total	31	12	6	11	5	6	6	77

Chi-Sq = 6.150, DF = 12, P-Value = 0.908

Expected Value of "B.T." in Defect type of the N/S
$E = (26 \times 31)/77 = 10.47$

Chi-Square = $(O-E)/E^2$

Chi-Square = $(13-10.47)^2/10.47$
            = 0.613

Since P Value > 0.05 ;
Reject Ha , Accept Ho

Hence Scrap in piston foundry is independent of shifts

The overall chi square came out to be 6.150 and value of p was 0.908 at 12 degree of freedoms. As it is higher than 0.005 (test was performed at 95% confidence level) so null hypothesis was accepted, so there existed no difference between the mean scrap value of morning, afternoon and night shift. It showed that this factor did not contribute substantially to scrap in the shop-floor and could be neglected.

Category-2

SSVs Selected: Die Coating Thickness

Tools Used: One-Way ANOVA

Under given temperature and time conditions, die had been made to run with various coating thicknesses (i.e. at 50 microns, 80 microns, 110 microns and 140 microns) separately. Basically impact of die coating thickness (inside the die) on bottom thickness variation and its associated scrap, needed to be checked. So for formulation of hypothesis, data was collected for conducting one-way ANOVA to four groups at respective die coating thicknesses (refer table 4.9). Data had four groups (more than two) so analysis of variance was more suitable as per guidelines.

Table-4.9 Data for One Way ANOVA

Null Hypothesis   Ho: $\mu1 = \mu2 = \mu3 = \mu4$ (Diecoating thickness do not effect Bottom thickness variation)
Actual Hypothesis Ha: $\mu1 \neq \mu2 \neq \mu3 \neq \mu4$ (Diecoating thickness effects Bottom thickness variation)

Die coating thickness 50μm	Die coating thickness 80μm	Die coating thickness 110μm	Die coating thickness 140μm
9.81	9.80	9.70	9.70
9.70	9.71	9.87	9.80
9.80	9.80	9.80	9.70
9.90	9.89	9.58	9.67
9.70	9.75	9.68	9.71
9.60	9.62	9.60	9.80
9.80	9.65	9.67	9.89
9.76	9.75	9.85	9.75
9.70	9.66	9.95	9.62
9.87	9.66	9.80	9.65
9.80	9.77	9.76	9.75
9.58	9.58	9.66	9.70
9.68	9.68	9.61	9.87
9.60	9.58	9.70	9.80
9.67	9.67	9.80	9.58
9.85	9.85	9.70	9.86
9.95	9.95	9.67	9.90
9.80	9.88	9.71	9.70
9.53	9.53	9.80	9.87
9.95	9.95	9.89	9.80

After giving input of above data, Minitab was used to compile ANOVA for all the four groups and p value came as 0.932 at 95% confidence level (as shown in figure 4.22). Again null hypothesis was accepted.

Figure-4.22 ANOVA for Different Coating Thickness Groups

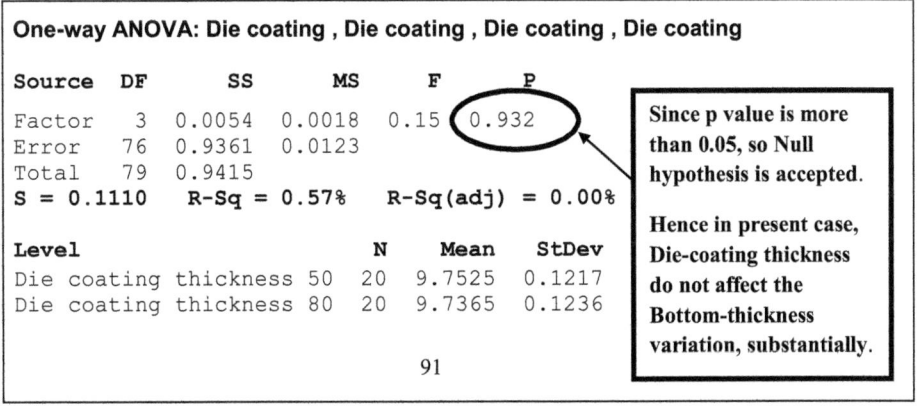

```
Die coating thickness 11   20   9.7400   0.1021
Die coating thickness 14   20   9.7560   0.0935
```

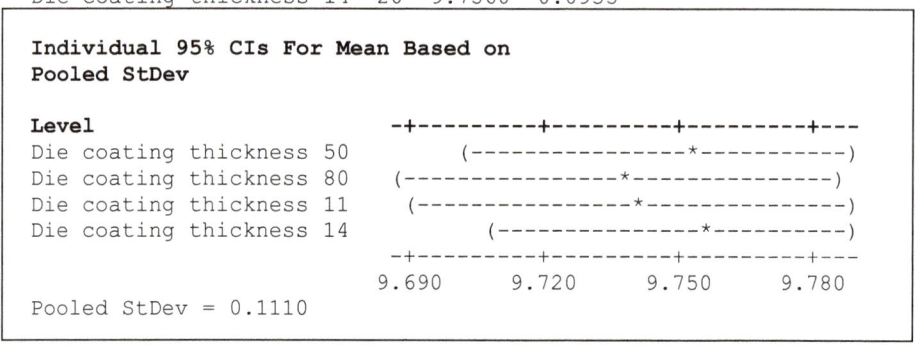

```
Individual 95% CIs For Mean Based on
Pooled StDev

Level                      -+---------+---------+---------+---
Die coating thickness 50              (----------------*-----------)
Die coating thickness 80   (----------------*----------------)
Die coating thickness 11     (----------------*----------------)
Die coating thickness 14              (---------------*----------)
                           -+---------+---------+---------+---
                          9.690     9.720     9.750     9.780
Pooled StDev = 0.1110
```

Standard deviation and mean of each group were postulated at 57% of R-square value and small variation was observed in them. Figure 4.23 represents normal probability plot, individual value plot and box plot in a sequence.

Figure-4.23 Analysis of Results

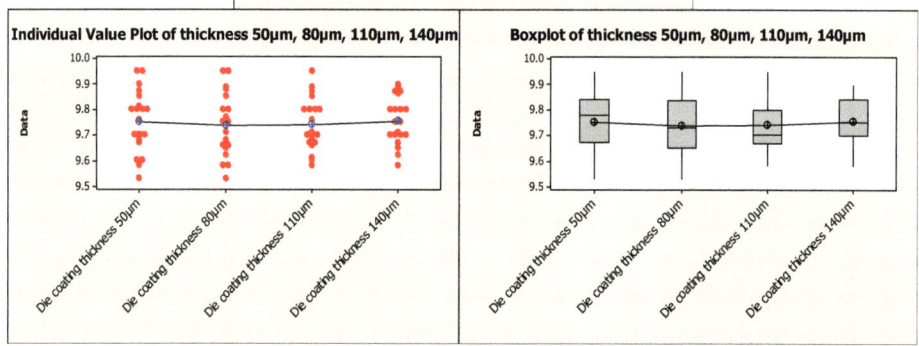

Data is normal as it is almost close to the straight line on normal plot. The mean of all the four groups is not varying much as it can be interpreted from the common line drawn by joining all means in individual value plot and box-plot graphs. Since this factor has also not been influencing the scrap much, hence it can be dropped during improve phase for optimization.

Category-3

SSVs Selected: In-Gate Design, Discharge of Cooling Water, Operator Skill

Tools Used: 2 Sample t-Test

This analysis contains two groups with each having sample less than 30, hence 2-sample t-test is best suited for such type of data populations (Dasgupta, T., 2003). For this analysis two piston dies were selected, one had existing runner riser volume (i.e. 250-260 cubic cm) and other was modified to the volume of runner riser up to 260-270 cubic cm (altered in-gate design). Six trials were taken (refer table 4.10) and each had considered at least 50 pistons to assess the actual effect of increase in volume to the percentage of casting scrap. The test was conducted at 10 degree of freedom and corresponding standard deviation was calculated as 0.0187. t-value was coming out to be 4.63 and probability value (p) was 0.001, which is less than 0.05 (value of alpha at 95% confidence value), so null hypothesis had been rejected and alternate hypothesis has been accepted for given two populations (refer figure 4.24). The graphical implications of t-Test results had been shown in figure 4.25 below.

Table-4.10 Data of In Gate Design

Analysis of Scrap(in %age) with variation in Volume of Riser & Runner

| Null Hypothesis | Ho: | $\mu1 = \mu2$ | Volume of riser & runner has no impact on casting scrap |
| Actual Hypothesis | Ha: | $\mu1 \neq \mu2$ | Volume of riser & runner has impact on casting scrap |

Trails	With Existing volume of Runner & Riser (240-250 cubic cm)	With Enlarged volume of Runner & Riser (260-270 cubic cm)
1	18%	15%
2	20%	16%
3	19%	14%
4	17%	12%
5	21%	13%
6	16%	11%

Casted 100 pistons per trail to assess the actual scrap in %age

Figure-4.24 Test for In Gate Design

Two-Sample T-Test and CI: With Existing volume of, With Enlarged volume of

Two-sample T for With Existing volume of Runner vs With Enlarged volume of Runner

	N	Mean	StDev	SE Mean
With Existing volume of	6	0.1850	0.0187	0.0076
With Enlarged volume of	6	0.1350	0.0187	0.0076

Difference = mu (With Existing volume of Runner) - mu (With Enlarged volume of Runner)

Estimate for difference: 0.0500

95% CI for difference: (0.0259, 0.0741)

T-Test of difference = 0 (vs not =): T-Value = 4.63

P-Value = 0.001

DF = 10

Conclusion: as P value is <0.05 So, Discard Ho and Accept Ha.

It implies, there is a relation between Runner & Riser Volume and casting scrap.

Figure 4.25 Representations of t-Test Results

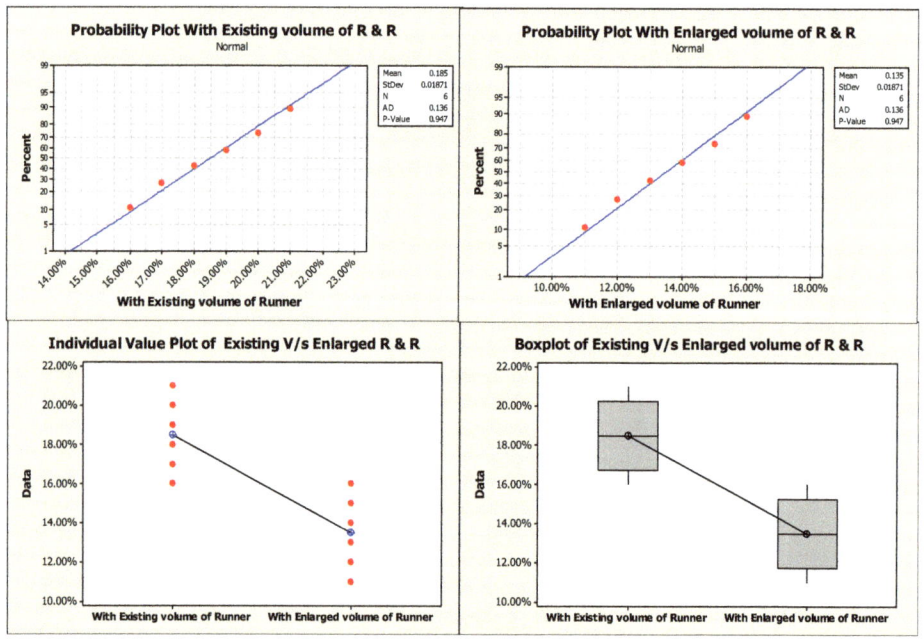

It was seen that with increase in runner and riser volume (or altering in gate design), casting defects were reducing because of proper feedback of molten metal after pouring and during solidification of casting in die moulds. First two charts in figure 4.25 are out comes of normality test done for data collected for corresponding values of scrap at mean 255 cubic cm and 265 cubic cm volume of runner and riser. Next two charts are individual value plot and box plot that are showing obvious difference in mean of two groups which can't be ignored.

Similarly t-test was used to analyse the effect of operator skill and discharge of cooling water on casting scrap. The p values were coming out to be 0.007 and 0.062 respectively for operator skill and discharge of cooling water cases. It indicates that skill level is a significant factor but discharge of cooling water has negligible effect, as p value is crossing 0.05 value.

Category-4
SSVs Selected: Alloy Temperature, Die Temperature and Delay Time
Tools Used: Multi-Regression

In the given case, three process parameters (namely; alloy temperature, die temperature and delay time) were considered as critical, as far as high scrap of piston castings was concerned. In order to verify their dependability on scrape rate, it was decided to perform multi-regression analysis. Data was collected randomly during regular die casting process in normal working conditions (refer Annexure 3). The average data for stoppage time and die temperature had been generated at certain values of temperature as shown in table 4.11.

Table-4.11 Input Data for Multi-Regression

Alloy Temperature	Stoppage (In Secs.)	Die Temperature	Total Pistons	Scrap	%age of Scrap
755	50	240	10	4	40.0
757	85	252	5	2	40.0
760	45	240	8	3	37.5
763	5	265	7	0	0.0
764	35	274	13	2	15.4
765	120	283	17	6	35.3
752	90	277	10	2	20.0
754	10	250	10	3	30.0

Before regression analysis, normality of data was checked by drawing normality plot and further dependability of each independent variable (alloy temp., die temp. and delay time) on dependable variable (scrap) was duly verified individually by drawing respective fitted line plots as described in figure 4.26.

Figure-4.26 Normality and Fitted Line Plots of Process Variables

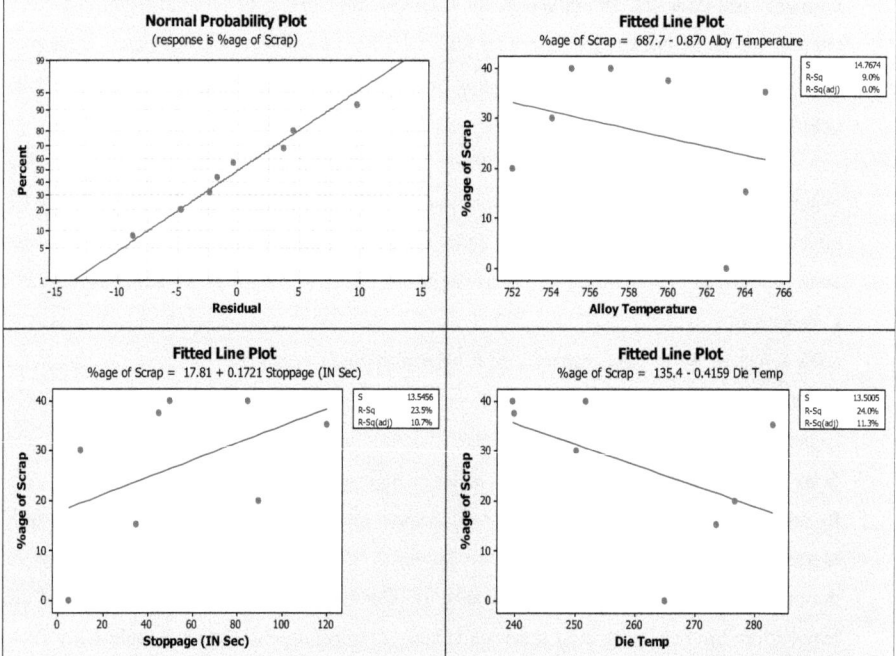

The negative slope of fitted line plot of alloy temperature and scrap reflects inverse relation in between them, similarly die temperature and scrap has also steep negative slope which means that as die temperature is increased from 240 to 280 degree Celsius, the scrap is decreased from 33% to 18% approximately. Delay time has positive slope, it means scrap is increased with increasing delay in casting process as highlighted. A regression equation was formulated for various Xs and Y (refer figure 4.27). The magnitude of coefficient of alloy temperature, die temperature and delay time have been indicated as the weightages of each independent variables on dependent variable respectively. Hence 0.7 of die temperature shows its maximum impact on scrap as compared to 0.3 of stoppage (delay time) and then 0.1 of alloy temperature etc. The p value of alloy temperature was more than 0.05 while p values of delay time and die temperature were less than this barrier.

Figure-4.27 Regression Analysis

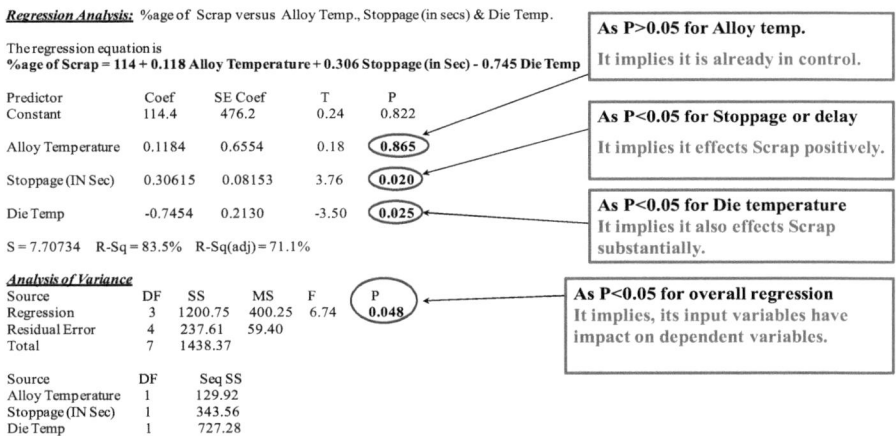

It clearly indicates that in the present case, the variation of alloy temperature is not much and does not contribute to scrap potentially as compared to other two variables. Lastly ANOVA was conducted to realize the variance in overall regression analysis of process parameters. The p value came out to be 0.048 which is less than 0.05, so the alternate hypothesis is accepted. The residue plots have been drawn to analyse the linear or non-linear relation of independent variables and dependent variable (refer figure 4.28).

Figure 4.28 Residue plots for Xs & Y

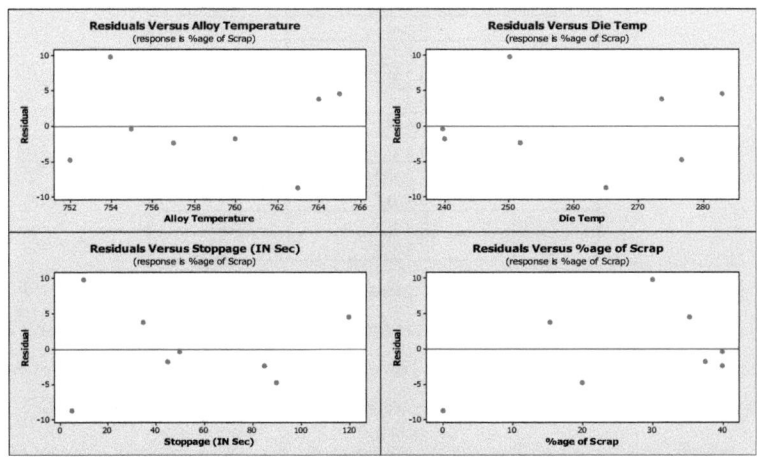

It shows that out of three, only two process variables have impact on casting scrap and must be taken care of. All the residue plots are perfectly random and none of the plot is presenting any trend or regular sequence. This shows that all the variables (Xs and y) are linearly correlated with one another and regression analysis performed previously seems to be authentic.

Category-5

SSVs Selected: Vaccume Pressure

Tools Used: Interaction Plot

To suck entrapped air and fumes from mould cavity, vaccume pressure is applied to air vents of die and it further reduces the defects like blow holes, porosity and cold lap in castings. It also ensures cleaning of air vents. There were mainly four dies and vaccume pressure could be applied in the range of 0.2 bar to 0.7 bar to each of them. Table 4.12 shows the scrap calculated in percentage during experimentation. At least 100 pistons were casted for each trial to find percentage of scrap feasibly.

Table-4.12 Interaction Plot Data

Die No.	Vaccume Pressure(in atm)	Scrap (in %age)
1	0.2-0.3	14
1	0.4-0.5	16
1	0.6-0.7	17
1	0.8-0.9	21
2	0.2-0.3	15
2	0.4-0.5	15
2	0.6-0.7	18
2	0.8-0.9	20
3	0.2-0.3	12
3	0.4-0.5	14
3	0.6-0.7	16
3	0.8-0.9	19
4	0.2-0.3	14
4	0.4-0.5	19
4	0.6-0.7	21
4	0.8-0.9	22

Figure 4.29 shows the effect of vaccume pressure on casting scrap with respect to different dies. It demonstrates that as the range of vaccume pressure is increased (from 0.7 to 0.2 bar), the scrap is reduced from 21% to 12%. The other chart indicates the relative sensitivity of dies as compared to one another. Die 3 seems to be highly responsive than others.

Figure-4.29 Interaction Plot among Three Parameters

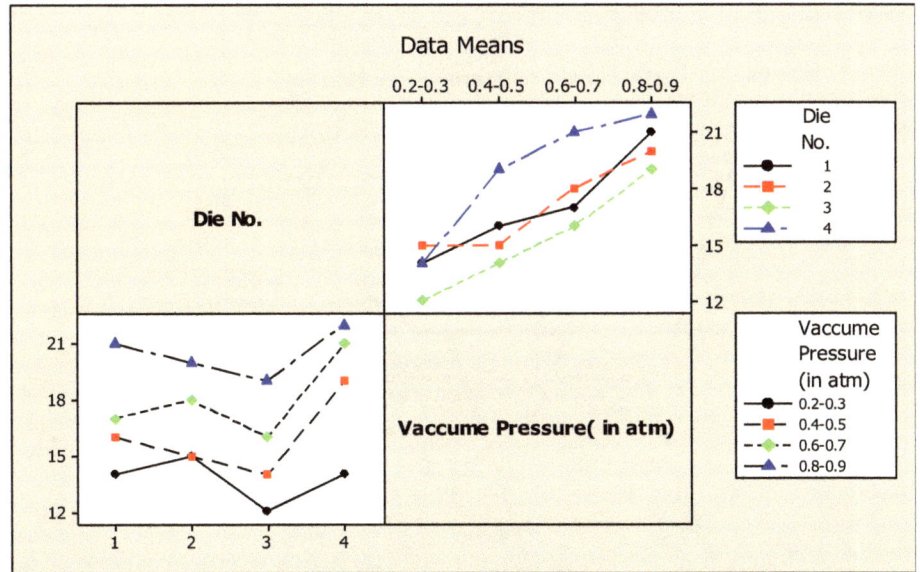

Category-6

SSVs Selected: Metal Sticking on Pins

Tools Used: Ishikawa Diagram

A fish bone analysis (Ishikawa diagram) has been used for analyzing the reasons of defective pin hole in castings (refer figure 4.30). It is one of the most compatible tool where brainstorming and reasoning are being used simultaneously. There are seven major reasons for this defect and each reason has several root causes. Next by brain storming all of these causes were classified into Control, Noise and Out of Control reasons for future course of action. It was brainstormed that metal sticking on die pins causes 75% of hole defects in castings and must be taken care of in improve phase. Actually metal used to stuck on die pins after 40 to 50 shots and then this unwanted stacked metal causes pin hole defect (internal scratching of holes in piston casting). At the end, poor cleaning procedure and dryness of die pins were shortlisted as the main reasons of metal sticking which causes defective holes in castings.

Figure-4.30 Fish Bone Analysis for Defectives Holes in Castings

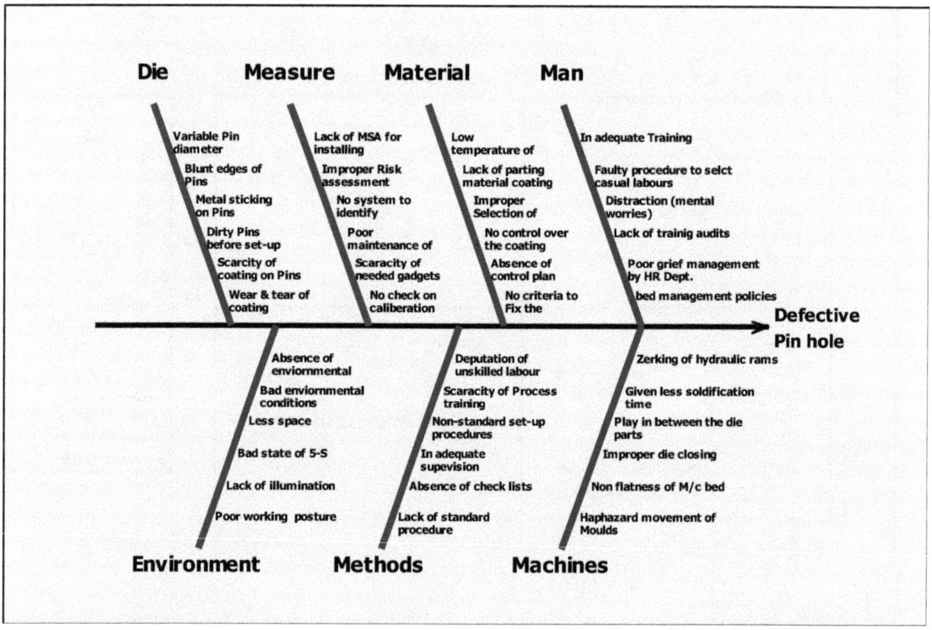

```
         Die          Measure        Material         Man
         |              |               |              |
    Variable Pin    Lack of MSA for  Low            In adequate Training
    diameter        installing       temperature of
    Blunt edges of  Improper Risk    Lack of parting Faulty procedure to selct
    Pins            assessment       material coating casual labours
    Metal sticking  No system to     Improper        Distraction (mental
    on Pins         identify         Selection of    worries)
    Dirty Pins      Poor             No control over Lack of trainig audits
    before set-up   maintenance of   the coating
    Scarcity of     Scarcity of      Absence of      Poor grief management
    coating on Pins needed gadgets   control plan    by HR Dept.
    Wear & tear of  No check on      No criteria to  bed management polices
    coating         caliberation     Fix the
    ───────────────────────────────────────────────────────────→ Defective
                                                                  Pin hole
    Absence of       Deputation of    Zerking of hydraulic rams
    enviornmental    unskilled labour
    Bad enviornmental Scaracity of Process Given less soldification
    conditions       training         time
    Less space       Non-standard set-up Play in between the die
                     procedures       parts
    Bad state of 5-S In adequate     Improper die closing
                     supevision
    Lack of illumination Absence of check lists Non flatness of M/c bed
    Poor working posture Lack of standard Haphazard movement of
                         procedure    Moulds
       Environment      Methods       Machines
```

Category-7

SSVs Selected: Machine's Dimensional In-accuracy

Tools Used: Why-Why Analysis

It was observed that due to dimensional in accuracy of machine bed and other parts, often head of piston casting got tilt (or bottom thickness variation came into picture). To eradicate this problem, why-why analysis had been performed (refer figure 4.31). The flow diagram of why-why analysis self explained the reasons of head tilt as shown above. Improper mating of die's bottom on outer moulds was the main responsible element for this. The other reason was the installation of un-checked die on machine by fitters but probability of this reason was quite low because of already implemented rigorous die- inspection procedure. By further questioning, two reasons of improper seating were found; first either by miss-alignment or play in between the ram of machine and die seat or other, miss-alignment of die's bottom tool with die seat. After next why, it was realized that first type of miss-alignment was due to mounting of bottom tool and catcher assembly on same ram and second type of miss-alignment was due to presence of less guiding pads in between the bottom tool and mould seat.

Figure 4.31 Why-Why Analysis

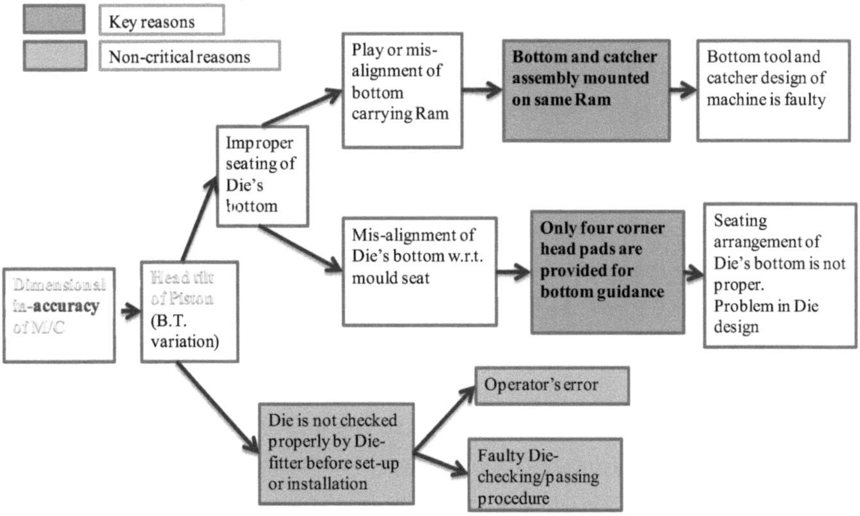

```
Key reasons
Non-critical reasons
```

Dimensional in-**accuracy** of M/C → Head tilt of Piston (B.T. variation) →

- Improper seating of Die's bottom →
  - Play or mis-alignment of bottom carrying Ram → **Bottom and catcher assembly mounted on same Ram** → Bottom tool and catcher design of machine is faulty
  - Mis-alignment of Die's bottom w.r.t. mould seat → **Only four corner head pads are provided for bottom guidance** → Seating arrangement of Die's bottom is not proper. Problem in Die design
- Die is not checked properly by Die-fitter before set-up or installation →
  - Operator's error
  - Faulty Die-checking/passing procedure

At the end it was deduced that the faulty design of machine's ram and availability of improper mating guides in piston die, were the root causes of head tilt or BT variation in castings.

4.6.3 Inferences from A-Phase

Major inferences from this phase are:

- After analyse phase, it became clear that out of eleven SSVs, only seven were actually responsible for high scrap.
- This phase analysed every factor with suitable statistical or non statistical techniques and found scrap was independent of shift, die coating thickness, discharge of cooling water and alloy temperature.
- The proposed criteria of analytical tools had helped to priorities the effort and helped to pick right tool or technique (qualitative or quantitative) for a given SSV analysis.
- This phase was vital as it analysed all SSVs completely and further highlighted only significant SSVs that should be taken into next phase for bringing expected breakthroughs in scrap reduction.

- By separating non-vital SSVs, this phase helped to decrease the net project-response time and efforts, appreciably.

4.7 Improve Phase

This is the most crucial phase and must be executed with great degree of seriousness. For necessary improvement, one can use various industrial engineering tools, lean tools and different optimization techniques separately or in an integrated way, whichever is more compatible (Raisinghani et al., 2005). In present study four casting parameters have been optimized through Design of Experiments (DoE). Some machine and die parts have been amended as per the principle of Poka-yoke for defect free operations. A kaizen has been performed to bring continuous improvement in pin cleaning activity and more efficient training schedules have also been incorporated for compensating concerned training needs of foundry workers and engineers. However, some pitfalls that can lead to a poorly designed improve phase are:

Lack of statistical knowledge: There are several reasons for the relatively low application of statistical methods in industry (Pfeifer et al., 2004). Managements in SMEs do not have the sufficient theoretical knowledge to see the potential of using statistical tools (Schroeder et al., 2008). In many cases, the companies and their employees even get frightened, when statistical tools are discussed. Most of companies also lack resources in the form of time and personnel (Perry and Barker, 2006).

Selection of wrong design: Many industrial engineers perform one-factor-at-a-time (OFAT) experiments to examine situations of process improvement and for problem-solving activities (Pheng and Hui, 2004). However, OFAT experiments can prove to be inefficient and unreliable and lead to false optimal conditions. Moreover, these often consist of "trial and error", relying on luck, intuition, guesswork and experience for their success. DoE is a powerful technique for discovering a set of important process or design variables and then assisting to determine at what levels these variables should be set/kept to optimize performance (Thomas and Barton, 2006).

Wrong planning: Careful planning can help to avoid problems that can occur during the execution of the experimental plan. For example personnel, equipment availability, funding and the mechanical aspects of system may affect ability to complete the experiment (Tang et al., 2007).

Lack of skilled manpower: Ignorance among workers on principles and procedures of various tools is one of reasons for failure of improvement phase (Van Den et al., 2006). The knowledge to fit the design or screen the designed experiments is vital to reap positive and favorable results, e.g. one should know how to apply full factorial DoE and should know when and where to apply fraction factorial design, if full factorial is not compatible with process constraints etc. (Rasis et al., 2002).

4.7.1 Proposed Framework for I-Phase

Although significant advances have taken place in metal casting simulation software, molding machinery, binder formulations and alloy development but still there is a need for process control capability to reduce the cost of production and improve profitability. Some foundries collect enormous amount of data, use statistical process control (SPC) and plot control charts and determine the process capability values and always insist that variability in the process parameters is a very big problem (Sarkar, 2007). It is agreed that variability is a problem. A process is considered as optimal only if there is no further scope for improvement in the cost of production left.

A two step framework has been formulated for improve phase as shown in figure 4.32. Aim of this phase has been defined by taking reference from output of analyse phase that seven critical SSVs need to be improved for possible results. Generally in practice, improvement can be achieved through five different ways namely; by optimization techniques, by using continuous improvement tools, by fool proofing, by providing required training or by innovation (Singh & Khanduja, 2011d). Tools available for each way have been quoted in second step.

Out of various optimization tools/techniques, DoE is a widely used tool in industry because:
- It can handle up to 200 factors and 40 responses at a time, which is not practical with the traditional statistical tools.
- The results are based on recognizing patterns in the surface diagrams rather than trying to fit one or more statistical distribution onto the data (Sahoo et al., 2008).
- It creates the process knowledge that can be documented for future use on the specific components (Rylander and Provost, 2006).

Designed experiments are often carried out in four phases: planning, screening (also called process characterization), optimization and verification. Based on this, an algorithm has been developed for systematic conduct of DoE in foundries (refer figure 4.33).

Figure 4.32 Proposed Framework for Improve Phase

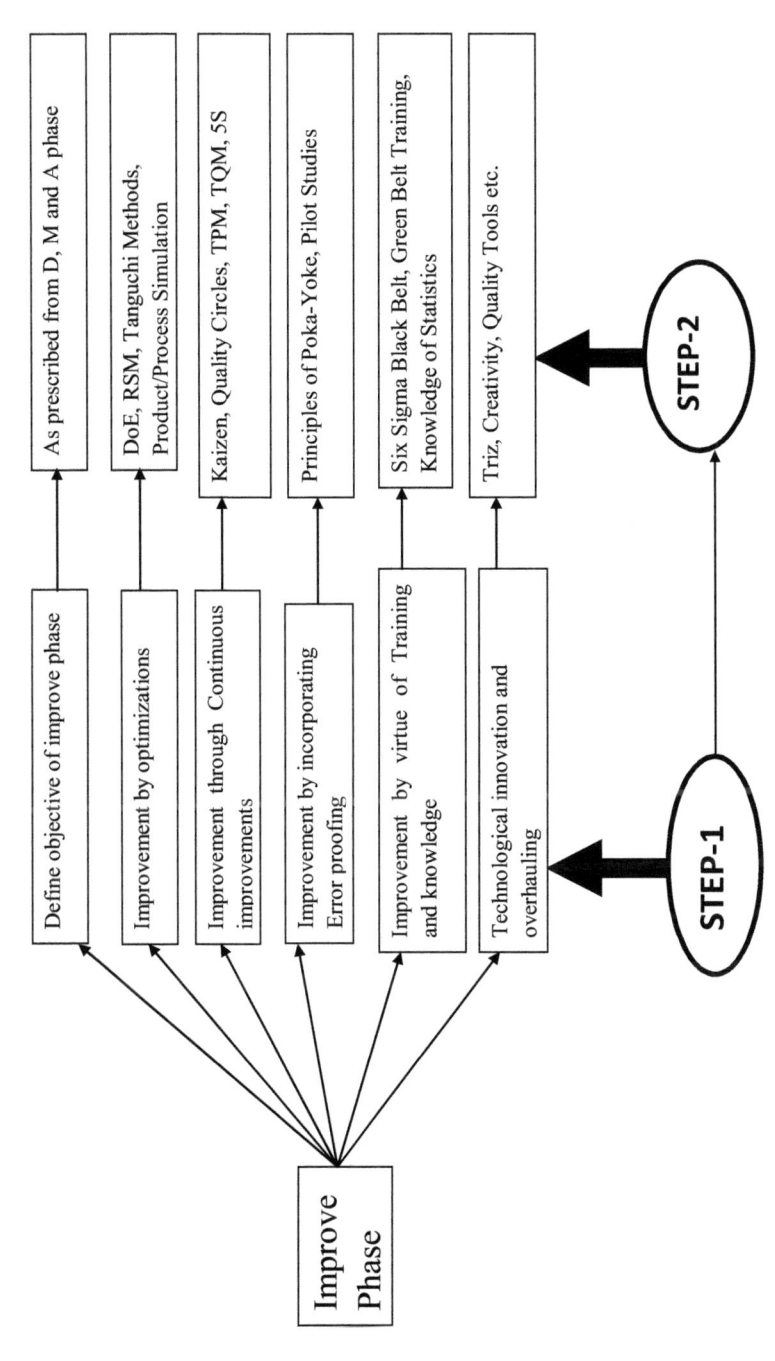

Different steps involved in DoE are:

Define the problem: In first step, identify the questions that are to be answered. At this step, define the goals (response) of the experiment. Shortlist the given critical to quality factors with their respective levels.

Select the design: During second step, selection of an experimental plan provides meaningful information. Full factorial or fractional factorial designs are falling under this category (Sharma, 2003). While in production experiments, small changes in selected variables of process have been made to find the optimum combination of variables. During response to surface experiments, only selected variables of process have been changed to generate an empirical map and the mathematical model of the response (Pheng and Hui, 2004).

Figure-4.33 Algorithm for an Integrated Methodology of DoE

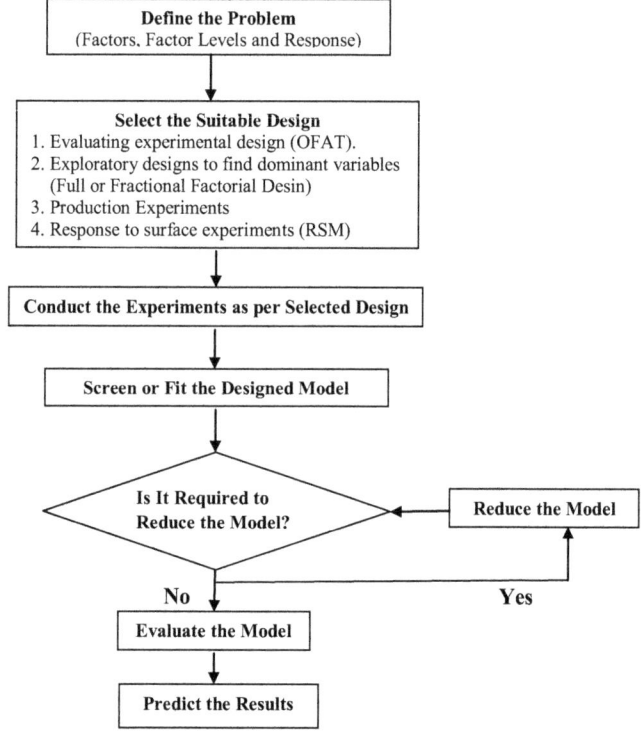

Conduct the experiment: The ordered sequence of the factor combinations (experimental conditions) is called the run order. It is usually a good idea to randomize the run order to lessen the effects of factors that are not included in the study, particularly effects that are time-dependent. When performing the experiment, measure the response (observation) at the predetermined settings of the experimental conditions. Each experimental condition that is employed to obtain a response measurement is a 'run'.

Screen and fit the design model: In many process development and manufacturing applications, potentially influential variables are numerous. Screening reduces the number of variables by identifying the key variables that affect product quality. This reduction allows focusing process improvement efforts on the really important variables, or the 'vital few'.

Reduce the model: After the "vital few" have been identified by screening, the "best" or optimal values for these experimental factors should be determined. Optimal factor values depend on the process objective. For example, it may require to maximizing the process yield or reducing product variability. The optimization methods available in Minitab include general full factorial designs (designs with more than two-levels), response surface designs, mixture designs, and Taguchi designs. Again the reduced model is further screened until it is not best fitted or reduces as possible.

Evaluate the model: Check the normality of data collected and residues are plotted against their fitted values. Main effect plot of each factor is created to discuss its impact on residue (out-put factor). Various Interaction plot have been plotted to evaluate the combined effect of two or more factors in groups on residue.

Predict the results: Last step is to draw conclusions from evaluated graphs and figures and give Optimum values to the given input parameters respectively.

4.7.2 Case Findings (I-Phase)

Tools used: *Full Factorial DoE, Poka-Yoke, Kaizen and On-Job Training (refer table 3.1)*

At the end of earlier phases seven SSVs were short listed for improvement and to tackle them, an improvement plan has been shown in table 4.13. It was decided to optimize the process parameters like; Volume of R&R, Die Temperature, Delay and Discharge of Cooling Water with the help of Full Factorial DoE. Problem of Machine's Dimensional In-Accuracy was suggested to handle with principles of Poka-Yoke. Similarly Metal Sticking on Die-Pins had been decided to check with continuous improvements through Kaizen. At last for human resource development, proper On-Job Training schedule had been drawn.

Table-4.13 Improvement Plan

Critical SSV's	Technique Used	Improvement Tool
Volume of Runner & Riser	Optimization	DOE
Die Temperature		
Delay Time		
Discharge of Cooling Water		
Machine's Dimensional In-Accuracy	Fool-Proofness	Poka-Yoke
Metal Sticking on Pins	Continuous Improvement	Kaizens
Operator Skill	Technical Awareness	On-Job Training

These SSVs have been put in four different categories, according to nature of improvement tool being used for bringing breakthroughs.

Category-1

SSVs Selected: Volume of R&R, Die Temperature, Delay Time and Discharge of Water
Tools Used: DoE

The present study has validated the application of DoE approach which has further assisted to achieve a reduction of scrap from 22% to 15%. To perform research or innovation in foundry environments, generally there is no criterion and experimentation at different values of input variables has been performed in hit and trail manner (Yang, 2004). With advent of DoE, not only the numbers of experiments have been reduced but also results can now be evaluated in a more organized way and their impact can be studied more deeply before doing necessary optimization of parameters for an elevated value of output (Gack and Robison, 2003). Pareto charts had measured shrinkage, porosity, blow holes, pin holes and bottom thickness variations in pistons, as the main reasons of scrap. The four process parameters, namely volume of runner & riser, die temperature, delay in casting and discharge of cooling water, have been selected for optimization through proposed algorithm of DoE.

Step-1: According to proposed methodology of DoE, first step is to define the problem of the case. Scrap reduction is the main problem which depends upon four selected casting process variables (factors) as explained in table 4.14.

Table-4.14 Description of Critical Factors with Respective Levels

Factors/ Levels	Die Temperature °C	Discharge of Water (Litre/ Minutes)	Delay (Seconds)	Volume of R&R (Cm³)
Low	250	7	60	260
High	330	10	180	285

Each factor is defined in terms of high and low levels and values of each level has been fixed after analyzing their respective affects on response (scrap).

Step-2: Next step is to select the design as per the given process constraints. To realize the affect of each factor or their interactional impact on scrap, 'Full factorial design' has been decided to use for optimizing selected process parameters (refer table 4.15).

Table-4.15 Full Factorial Design of Experiments

StdOrder	RunOrder	CenterPt	Blocks	Die Temp	Discharge of water	Delay	Volume of R&R
3	1	1	1	-1	1	-1	-1
27	2	1	1	-1	1	-1	1
6	3	1	1	1	-1	1	-1
4	4	1	1	1	1	-1	-1
1	5	1	1	-1	-1	-1	-1
29	6	1	1	-1	-1	1	1
23	7	1	1	-1	1	1	-1
13	8	1	1	-1	-1	1	1
11	9	1	1	-1	1	-1	1
14	10	1	1	1	-1	1	1
9	11	1	1	-1	-1	-1	1
19	12	1	1	-1	1	-1	-1
24	13	1	1	1	1	1	-1
22	14	1	1	1	-1	1	-1
28	15	1	1	1	1	-1	1
25	16	1	1	-1	-1	-1	1
8	17	1	1	1	1	1	-1
16	18	1	1	1	1	1	1
12	19	1	1	1	1	-1	1
26	20	1	1	1	-1	-1	1
32	21	1	1	1	1	1	1
17	22	1	1	-1	-1	-1	-1
18	23	1	1	1	-1	-1	-1
2	24	1	1	1	-1	-1	-1
15	25	1	1	-1	1	1	1
20	26	1	1	1	1	-1	-1
10	27	1	1	1	-1	-1	1
5	28	1	1	-1	-1	1	-1
31	29	1	1	-1	1	1	1
7	30	1	1	-1	1	1	-1
30	31	1	1	1	-1	1	1
21	32	1	1	-1	-1	1	-1

- Full Factorial design
- Number of factors = 4
- Factor levels = 2
- Number of replications for corner points = 2
- Base for random data generator = 9
- Number of centre points = 0
- Blocks =1
- Runs = 32 (16 each)
- All runs are free from aliasing

No blocking is used and experiments were replicated twice for suitable accuracy. So it requires 2^4 or 16 experiments and random repetition of these 16 experiments again (as replica is 2) or 32 runs have been needed to perform, for generating feasible impacts of each factor over the response.

Step-3: The third step is to execute the experiments as per the run order of the created design. The respective scrap value in percentage was calculated for each run as cited in table 4.16. For each run values of factors, around hundred pistons were casted and out of these pistons, scrap was separated and percentage of scrap for each run was computed.

Table-4.16 Execution of Designed Experiments

StdOrder	RunOrder	CenterPt	Blocks	Die Temp	Discharge of water	Delay	Volume of R&R	Scrap (%)
3	1	1	1	250	10	60	260	14
27	2	1	1	250	10	60	285	10
6	3	1	1	330	7	180	260	18
4	4	1	1	330	10	60	260	17
1	5	1	1	250	7	60	260	15
29	6	1	1	250	7	180	285	11
23	7	1	1	250	10	180	260	12
13	8	1	1	250	7	180	285	11
11	9	1	1	250	10	60	285	9
14	10	1	1	330	7	180	285	16
9	11	1	1	250	7	60	285	10
19	12	1	1	250	10	60	260	13
24	13	1	1	330	10	180	260	18
22	14	1	1	330	7	180	260	20
28	15	1	1	330	10	60	285	15
25	16	1	1	250	7	60	285	9
8	17	1	1	330	10	180	260	18
16	18	1	1	330	10	180	285	12
12	19	1	1	330	10	60	285	13
26	20	1	1	330	7	60	285	15
32	21	1	1	330	10	180	285	13
17	22	1	1	250	7	60	260	16
18	23	1	1	330	7	60	260	19
2	24	1	1	330	7	60	260	18
15	25	1	1	250	10	180	285	9
20	26	1	1	330	10	60	260	17
10	27	1	1	330	7	60	285	15
5	28	1	1	250	7	180	260	17
31	29	1	1	250	10	180	285	10
7	30	1	1	250	10	180	260	14
30	31	1	1	330	7	180	285	16
21	32	1	1	250	7	180	260	17

Before conducting experiments some technical changes were incorporated in the casting machine and piston die e.g. an electric heater was permanently installed around the die and further connected with electronic auto cut arrangement, so that die's temperature could be controlled with in fixed values. Similarly an appropriate Rotameter (a discharge measuring device) was installed on machine to have a close check on flow of cooling water.

Step-4: In order to screen the experiments, the above data was executed with Minitab software and analysed. The matrix of experiments with results is shown in figure 4.34.

Figure-4.34 DoE Analysis of an Orthogonal Array

Estimated Effects and Coefficients for Scrap (%) (coded units)

Term	Effect	Coef	SE Coef	T	P
Constant		14.281	0.1362	104.84	0.000
Die Temp	3.937	1.969	0.1362	14.45	0.002
Discharge of water	-1.813	-0.906	0.1362	-6.65	0.005
Delay	0.438	0.219	0.1362	1.61	0.128
Volume of R&R	-4.313	-2.156	0.1362	-15.83	0.000
Die Temp*Discharge of water	0.062	0.031	0.1362	0.23	0.821
Die Temp*Delay	-0.187	-0.094	0.1362	-0.69	0.501
Die Temp*Volume of R&R	0.563	0.281	0.1362	2.06	0.056
Discharge of water*Delay	-0.687	-0.344	0.1362	-2.52	0.023
Discharge of water*Volume of R&R	0.312	0.156	0.1362	1.15	0.268
Delay*Volume of R&R	-0.187	-0.094	0.1362	-0.69	0.501
Die Temp*Discharge of water*Delay	0.187	0.094	0.1362	0.69	0.501
Die Temp*Discharge of water*Volume of R&R	-0.813	-0.406	0.1362	-2.98	0.009
Die Temp*Delay*Volume of R&R	-0.313	-0.156	0.1362	-1.15	0.268
Discharge of water*Delay*Volume of R&R	-0.313	-0.156	0.1362	-1.15	0.278
Die Temp*Discharge of water*Delay*Volume of R&R	-0.437	-0.219	0.1362	-1.61	0.128

S = 0.770552   PRESS = 38

R-Sq = 97.09%   R-Sq(pred) = 88.36%   R-Sq(adj) = 94.36%

(■ *Readings in red are statistically Significant;* ■ *Readings in green are Non-Significant)*

At 95% confidence level, the p value of three factors namely; die temperature, discharge and volume of R&R was lower than 0.05 and hence statistically critical as far as casting scrap was concerned. The fourth factor called 'Delay' was generated as non-critical factor as corresponding p value was coming out to be more than 0.05. The value of the coefficient of each factor represents the affect of each on response (scrap). It implies from above analysis that volume of R&R has more effect on scrap than die-temperature and discharge of cooling water. In figure 4.35 over all ANOVA has further made it clear that main effects and three way interactions had more impact on scrap than 2 way interactions (as concluded from p values). The term 'main effect' is defined as the individual impact of each factor on scrap and similarly, two way or three way interaction are just various combinations of factors either in two or in three numbers simultaneously. If the p values are not generated then one can think to randomize the experiment again or to revise the repeatability of experiments or think about the necessary blocking or at last can go for fraction factorial DoE rather than full factorial.

Figure-4.35 ANOVA Applied over DoE Model

Analysis of Variance for Scrap (%) (coded units)						
Source	DF	Seq SS	Adj SS	Adj MS	F	P
Main Effects	4	300.625	300.625	75.1563	126.58	0.001
2-Way Interactions	6	7.687	7.687	1.2812	2.16	0.103
3-Way Interactions	4	7.125	7.125	1.7813	3.00	0.050
4-Way Interactions	1	1.531	1.531	1.5312	2.58	0.128
Residual Error	16	9.500	9.500	0.5937		
Pure Error	16	9.500	9.500	0.5938		
Total	31	326.469				

(■ *Readings in red are statistically Significant;* ■ *Readings in green are Non-Significant*)

In present case figure 4.36 shows normal plot of all possible factors and clearly highlights A, B, D and ABD as more critical factors and factor interactions, which are affecting overall casting scrap substantially, as these are dotted in red and seems to be far away from the normal plot line of standardized effects.

Figure-4.36 Normal Plot of Effects (to foreground CTQs)

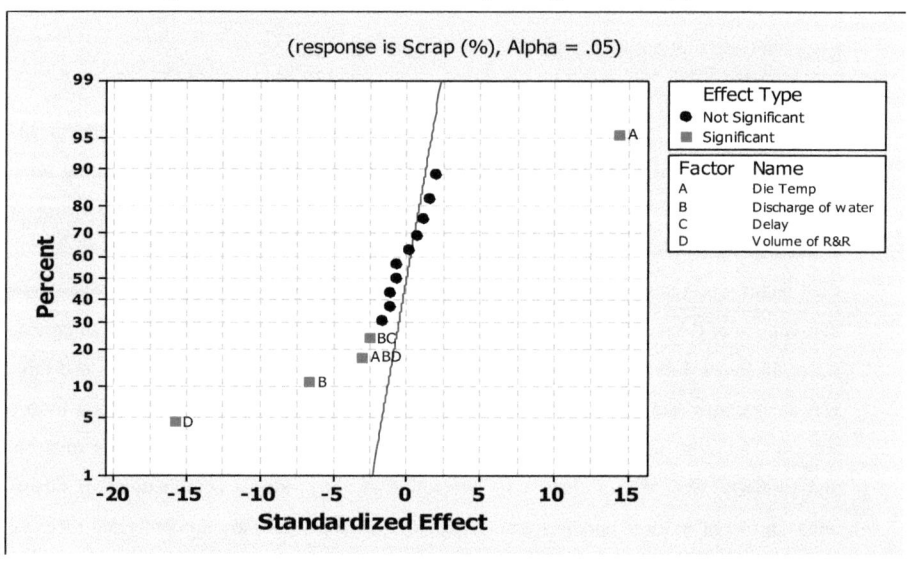

The graph is drawn at 95% confidence level and also shows relative percentage of significance of every factor, as the question of low scrap in casting process is raised.

Step-5: Now as per the methodology proposed, it is the time to reduce the experimental model. The figure 4.36 has already signaled the various critical factors and their critical interactions that have affected the scrap. All the other non-significant factors and interactions have been removed from the orthogonal matrix, as shown in figure 4.37.

Figure-4.37 Analysis of Reduced Model

Estimated Effects and Coefficients for Scrap (%) (coded units)					
Term	Effect	Coef	SE Coef	T	P
Constant		14.281	0.1448	98.61	0.000
Die Temp	3.937	1.969	0.1448	13.59	0.001
Discharge of water	-1.813	-0.906	0.1448	-6.26	0.004
Delay	0.438	0.219	0.1448	1.51	0.143
Volume of R&R	-4.313	-2.156	0.1448	-14.89	0.000
Discharge of water*Delay	-0.687	-0.344	0.1448	-2.37	0.026
Die Temp*Discharge of water* Volume of R&R	-0.813	-0.406	0.1448	-2.80	0.010
S = 0.819298 PRESS = 27.4944					
R-Sq = 94.86% R-Sq(pred) = 91.58% R-Sq(adj) = 93.63%					

(■ Readings in red are statistically Significant;   ■ Readings in green are Non-Significant)

The results of reduced or focused DoE shows that besides main factors and three way interactions, two way interactions have also been effecting scrap or response factor. It is a more focused effort to understand the effect of critical process factors and their respective interatcions (refer figure 4.38).

Figure-4.38 ANOVA Applied on Reduced Model

Analysis of Variance for Scrap (%) (coded units)						
Source	DF	Seq SS	Adj SS	Adj MS	F	P
Main Effects	4	300.625	300.625	75.1563	111.96	0.000
2-Way Interactions	1	3.781	3.781	3.7812	5.63	0.026
3-Way Interactions	1	5.281	5.281	5.2813	7.87	0.010
Residual Error	25	16.781	16.781	0.6713		
Lack of Fit	9	7.281	7.281	0.8090	1.36	0.282
Pure Error	16	9.500	9.500	0.5938		
Total	31	326.469				

Unusual Observations for Scrap (%)						
Obs	StdOrder	Scrap (%)	Fit	SE Fit	Residual	St Resid
3	6	18.0000	19.7722	0.3737	-1.7722	-2.43R
15	28	15.0000	13.5036	0.4166	1.4964	2.12R
16	25	9.0000	10.5529	0.3599	-1.5529	-2.11R

R denotes an observation with a large standardized residual.

Estimated Coefficients for Scrap (%) using data in uncoded units

Term	Coef
Constant	57.4014
Die Temp	0.0326925
Discharge of water	-0.709672
Delay	0.0361111
Volume of R&R	-0.190088
Discharge of water*Delay	-0.00381944
Die Temp*Discharge of water* Volume of R&R	7.13494E-06

The p value for two way interactions was 0.026 which is less than 0.05 and hence it was critical which are earlier ignored. Further in two way interactions, only combination of B (Discahrge) and C (Delay) factor is more critical rather than any other combination. The coffecient values are closer to true effects in reduced model analysis with respect to earlier full model analysis. Finally the optimum equation in between the independent and dependent variables was generated from DoE, as shown below:

Scrap (y) = **57.4 + 0.03** (A) **- 0.71** (B) **+ 0.03** (C) **- 0.19** (D) **- 0.003** (B * C) **+ 7.13** (A * B * D)

Scrap (y) = **57.4 + 0.03** (Die Temp) **- 0.71** (Discharge of water) **+ 0.03** (Delay) **- 0.19** (Volume of R&R) **- 0.003** (Discharge of water * Delay) **+ 7.13** (Die temp. * Discharge * Volume of R&R)

To analyse the relative impact (positive or negative) of each factor and respective interactions, a Pareto chart was drawn (refer figure 4.39). This figure highlights that factor B has effect even lower than the threshold effect value of 2.06 because it was a non significant factor as indicated previously.

Figure-4.39 Relative Effects of Factors

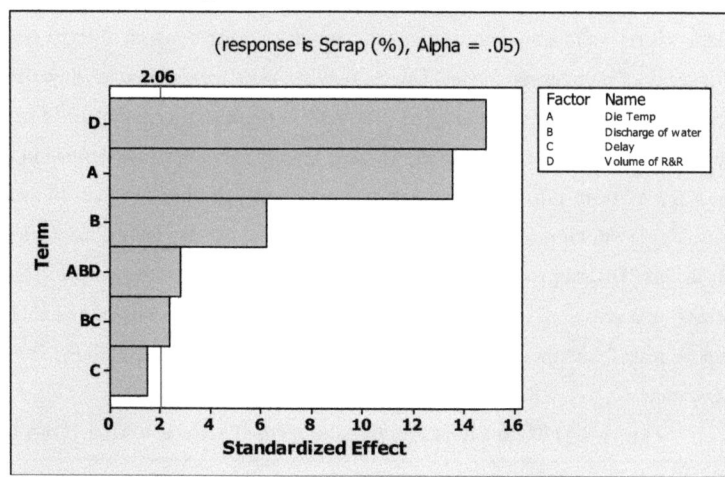

Step-6: This step evaluates DoE and its results. Four pack chart (refer figure 4.40) has been plotted showing normal probability plot, residue analysis with its fitted values, frequency and observation orders.

Figure 4.40 Four Pack Chart of Fitted DoE

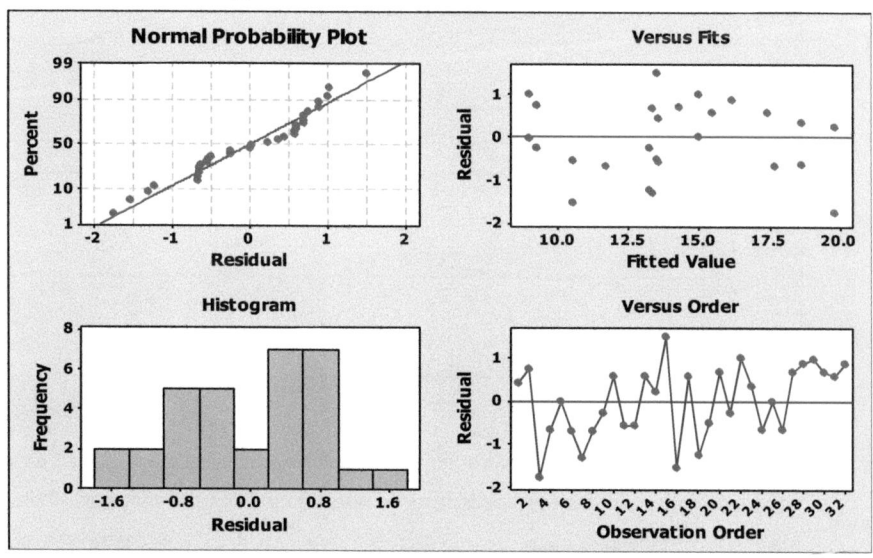

The collected data during experimentation seemed to be normal from probability plot and residue charts were also not making any trend or pattern which directly indicates the authenticity of the reduced model. The main effect plots were drawn to show the individual effect of each factor on response. The factors (4 in numbers) were plotted in between their respective two levels and against scrap to show their impact clearly, as defined in figure 4.41. The slope of main effect line in each plot represents the high impact of that factor on response, like 'Volume of R&R' and 'Die Temperature' are two factors having slope steeper than factors 'Discharge of water' and 'Delay'. It also evaluates the trend of response with increase or decrease of each factor value, independently. For example the value of casting scrap decreases with increase in volume of R&R while it decreases with the increase of die temperature etc.

Figure-4.41 Main Effect Plot by taking One Factor at a Time (OFAT)

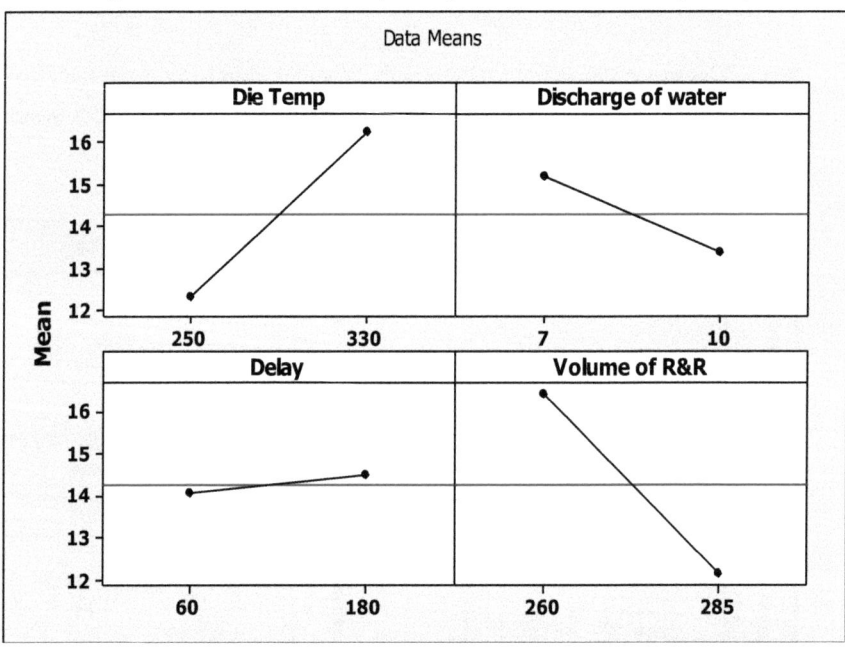

Delay has smallest slope as it was a non significant factor or having low impact on scrap value and finally, with increase in discharge of water from 7 liter per minute to 10 liter per minute the corresponding value of scrap decreases from 15% to 13%, as described in second plot of above figure. Similarly effect plots for two way interactions have also been drawn and presented in figure 4.42.

Figure-4.42 Two Way Interaction Plots

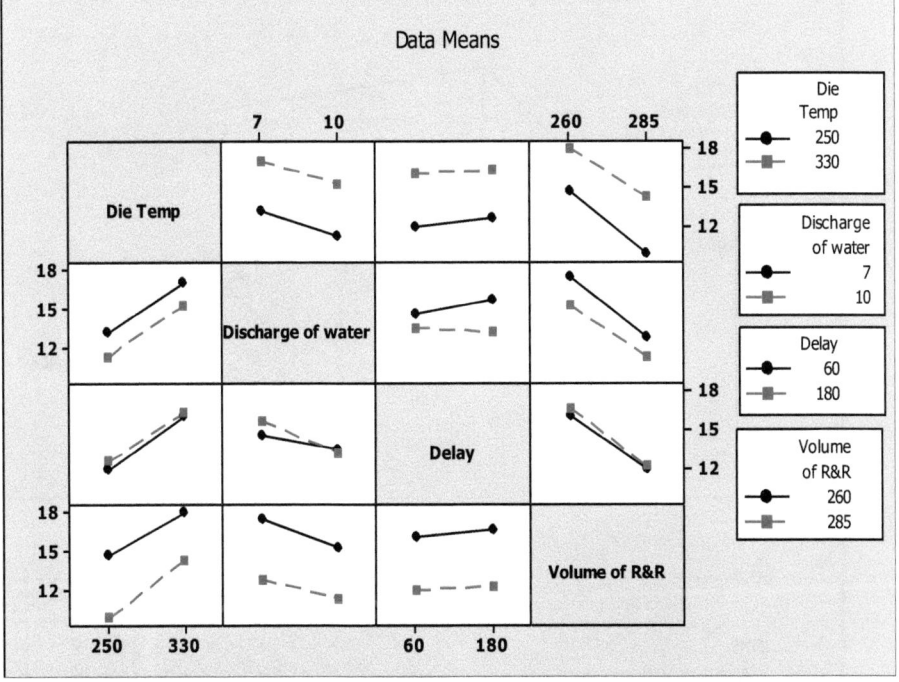

Graphically it is inferred that BC or combined effect of discharge of cooling water and delay has vital impact on scrap as two effects are crossing with each other. Similarly other interactions of delay with volume of R&R and die temperature are also affecting scrap up to certain level but it is worth to point here that the main effect of delay was coming out to be less significant. Out of three two way interactions of delay, the combination with discharge (C) was more effective, so it was decided to draw a surface plot of scrap versus two way interaction BC (refer figure 4.43).

The green and blue area of contour plot is just a side projection of surface plot which shows the regions of discharge and delay values corresponding to 10 to 12% scrap and 14 to 16% of scrap respectively. This plot has not only shown the effect of BC interaction but also provides the independent values of B and C for possible reduction of scrap, in the given

Figure-4.43 Surface and Contour Plot for Two Way Interaction (BC)

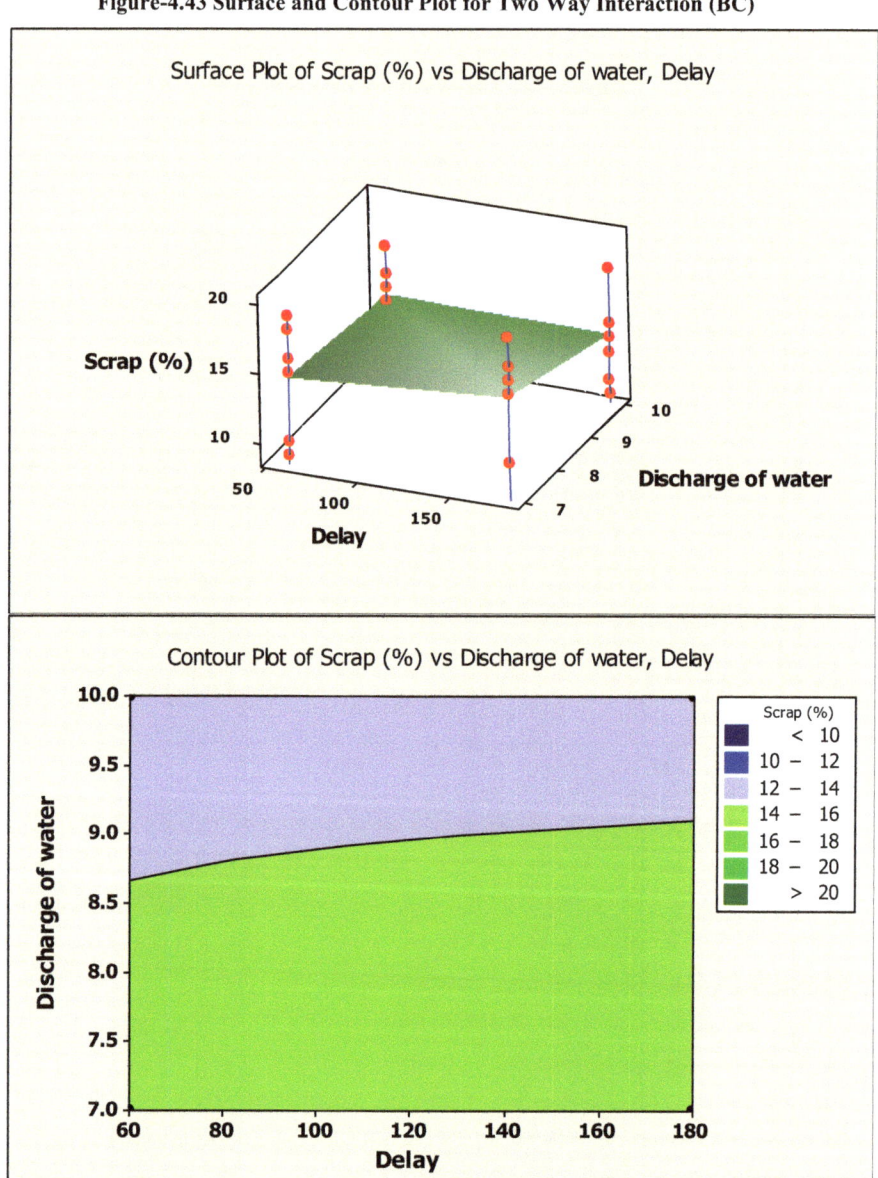

casting scenario. At the end of evaluation step, three way interactions have also been analysed, because only ABD interaction was significant, so a cube plot has been generated for analysis of this fact. In figure 4.44, cube plot has made it clear that at higher value of A (Volume of R&R), lower value of B (Die temperature) and at higher value of C (discharge of cooling water) the overall scrap will be minimum (9.5% only). It is less as compared to any other combination of three factors (A, B and D).

Figure-4.44 Cube Plot for Three Way Interaction (ABD)

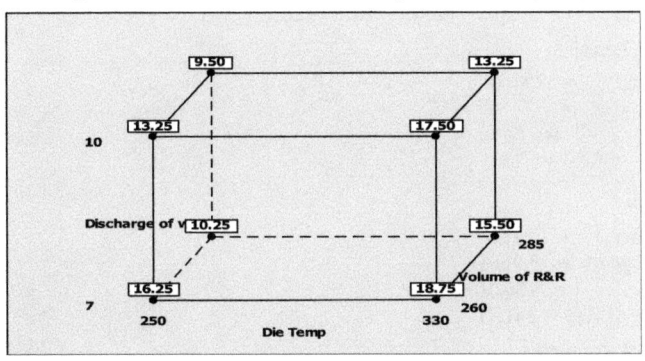

Step-7: In the last step results were predicted for the defined problem. To optimize the response, Response Optimizer or Overlaid Contour Plot was used to obtain a numerical and graphical analysis (refer figure 4.45).

Figure-4.45 Optimization of Process Parameters

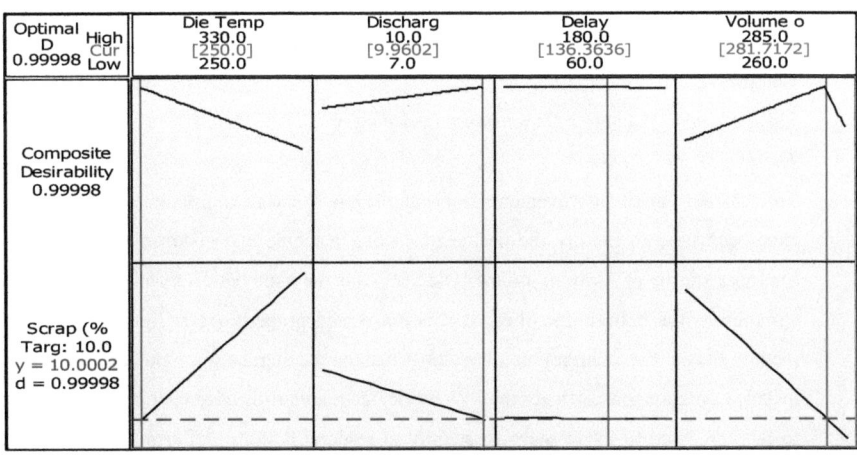

All the factors and their lower levels were quoted in optimizer. The existing value and the required value of the response were fed with its possible upper and lower values. The optimizer in Minitab software has given results at 99% desirability, as shown in figure 4.46.

Figure-4.46 Results from Response Optimizer

```
Response Optimization
Parameters
                Goal    Lower  Target  Upper  Weight  Import
Scrap (%)       Target    8      10     20      1       1
Starting Point
Die Temp     = 250
Discharge of =   7
Delay        =  60
Volume of R& = 260
Solution
Die Temp     =   250
Discharge of =   9.96019
Delay        =   136.364
Volume of R& =   281.717
Predicted Responses
Scrap (%) = 10.0002, desirability = 0.999982
Composite Desirability = 0.999982
```

These values for above critical parameters have been optimized to restrict casting scrap up to 10% only but in actual practice it was found that due to some un avoidable reasons or process constraints, scrap is actually reduced approximately from 22% to 15%.

Category-2

SSVs Selected: Machine's Dimensional In-accuracy

Tools Used: Poka-Yoke

To inculcate further improvements in casting scrap, Poka-Yoke principles of fool proofness have been used to modify the design of casting machine and piston die, reasonably. The improper seating of 'Bottom' on the 'Die Seat' was the main reason for high scrap due to BT variation as the bottom got tilted, if it is not seated properly (refer figure 4.47). The first picture shows the configution of casting mahine having bottom tool and catcher arms installed on common vertical ram with 90 degree angular dispalcement on same plane. Due to this ram used to move more frequently to respond bottom tool arm and catcher arm in every shot of casting. Because of these frequent movements, improper seating of bottom tool on die parts often took place. To sort out this, some changes were made in the design of casting machine.

Figure-4.47 Improvement in Machine Design

Bottom tool was mounted on separate hydraulic ram through overhead beam of machine while catcher assembly was put on same sidewise ram. Due to this bifurcation both movements could be executed by respective rams and hence probability of play among parts was halved with time. Other reason of BT variation was due to misalignment of bottom tool with respect to its pads on upper half of outer moulds. There are four pads provided on top of moulds for seating (refer figure 4.48). If any one of the pad did not fit properly then bottom got tilted and casting was scraped because of BT variation.

Figure 4.48 Modification in Die Design

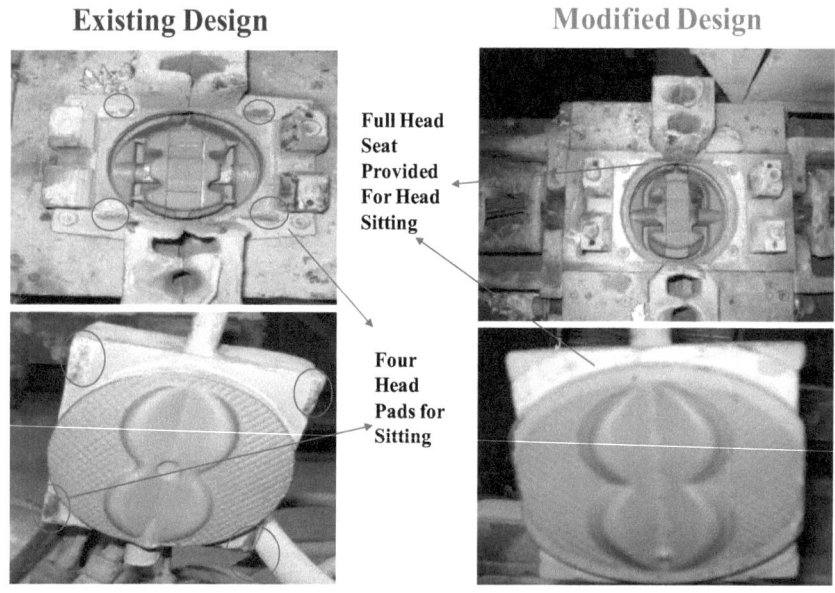

To rectify these, design of die parts was chaged strategically. Full head seat groove and pad was provided on bottom tool and upper part of moulds respectively. With this the chance of misalignment was eliminated completely and overall mating of die parts became more fool proof, reducing scrap from 2% to 2.5%.

Category-3

SSVs Selected: Metal Sticking on Pins

Tools Used: Kaizen

Actually after 65 to 70 shots of castings, the alloy metal starts sticking to the pins of die due to reasons like; rough surface of pins, low temperature of pins, improper cooling of pins while soldification of piston casting etc. Metal sticking on pins used to drag alloy from pin holes of castings as soon as outer moulds were open for ejection of casted pistons. Stuck metal was removed with needle file manually and as such no standard procedure for pin cleaning was there. So, an appropriate procedure for cleaning of pins was chalked out (refer figure 4.49).

Figure 4.49 Contineous Improvement in Pin-Cleaning Operation

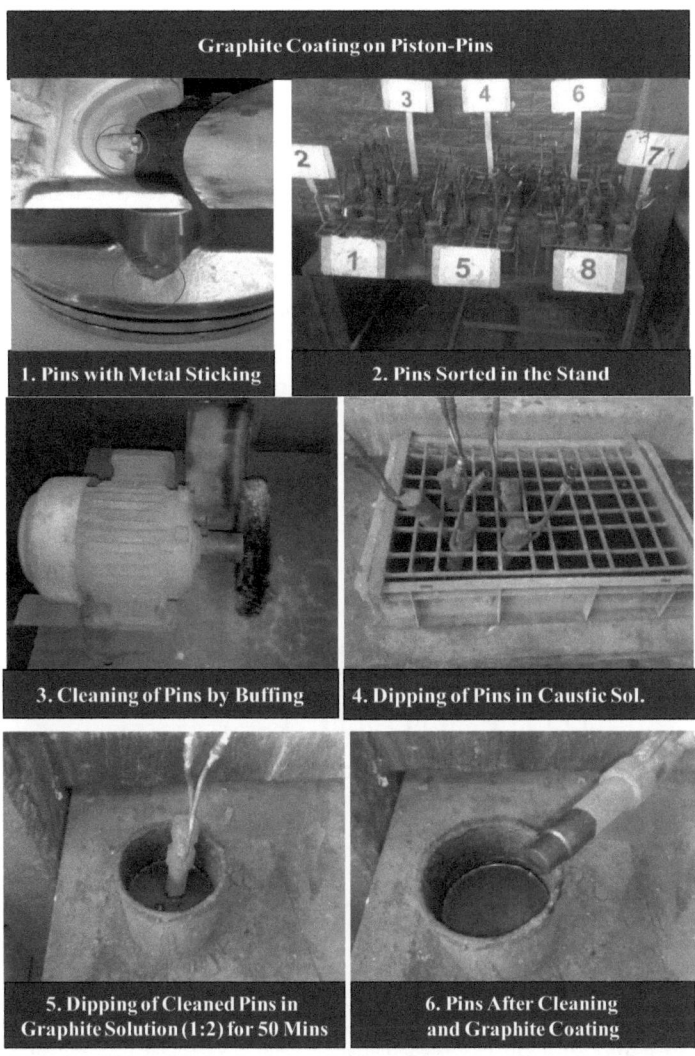

This involved use of buffer motor, caustic treatment for organic and inorganic stains and use of graphite solution to create a lubricated layer of carbon around the pins that could provide protection against metal sticking and hence provide better surface finish for pin holes in the casted pistons. Test trials showed that pins prepared with improved procedure had a life of 300 to 350 shots which means five times more pin's life. So this contineous improvement in pin cleaning procedure, not only reduced the scrap by 2% but also decreased the breakdowns.

Category-4

SSVs Selected: Operator Skill

Tools Used: On-Job Training

In foundry, there were no training schedules for training of workers. To prepare proper training schedule, various training needs were found out. Human capabilities were captured by making suitable training matrices (refer figure 4.50). This matrix can tell us the knowledge and skill level of workers and can predict the type of training needed. Based upon these matrices, an appropriate training schedules was prepared for workers (skilled and un-skilled), engineers and supervisors (refer Annexure 4 and Annexure 5). After such training, scrap got reduced around 1% and it is remarkable for a foundry SME.

Figure-4.50 Training Matrix

Cell:	H-273 Opistons	DCM20	DCM9	SELF INSPECTION	TEMPERATURE CHECKING	WASHING MAN	LINE HELPER/RING SHOT BLASTING	CORE MAKING	DIE FITTER	Master Trainer:
Shift:	All									Girish Kumar / Daljeet Singh
Supervisor	DPS/SM/MIJ									
Status Date:	30.08.2010									Comments, Training Plans
51051	TARLOK SINGH			U	☐		L	L	☐	L	To Be Trained for Temperature Checking of Holding Furnaces by 30 Sept.
51267	GURBAZ SINGH	U	U	L	L	L		L	L	To be trained to shot blast and clean the die with shot blasting machine by 4 oct.
51807	PAWAN KUMAR	U	☐	U		U	☐	U	L	to be trained to see temp by pyrometer by 30 sept.
51150	SUNIL KUMAR			☐	L		—	L	L	To be Trained on Full Self Inspection by 30 Sept.
51751	JASSA SINGH			☐	L		—	L	L	To be trained to do degassing and de slagging the molten metal by oct 5.
56771	GURMEET SINGH			U	U	L	—	L	U	L	Trained the working on die casting machine 20 by 4 oct.
Empty Square	No training completed to date.	—	Knows and applies all job instructions and safety rules …		U	and produces the quantity required by production targets.				
			L	and ensures quality according to work station job instructions.		☐	and is capable of training other team members.			

Financially, it was estimated that approximate saving of Rs. 8.78 lakhs per annum was achieved by reducing the scrap of H-273 pistons from around 22% to 10%. The whole improvement history of scrap has been plotted in figure 4.51 and detailed analysis of financial gains achieved can be described as below:

Empirical Analysis of H-273 pistons:

Average scrap saved per month = 400 pistons (Approximately)

Cost per piston: Material cost = Rs 45.75/-

Labour cost = Rs 27.45/-

Energy cost = Rs 82.35/-

Infrastructure cost = Rs 18.30

Remelting cost = Rs 9.15

Net cost of a piston = Rs 183/-

Scrap cost saved per month = 400*183 = Rs 73,200/-

Scrap cost saved per year = 73200*12 = Rs 8,78,400/-

Figure-4.51: Scrap Trend (from 1st July to 31st December 2010)

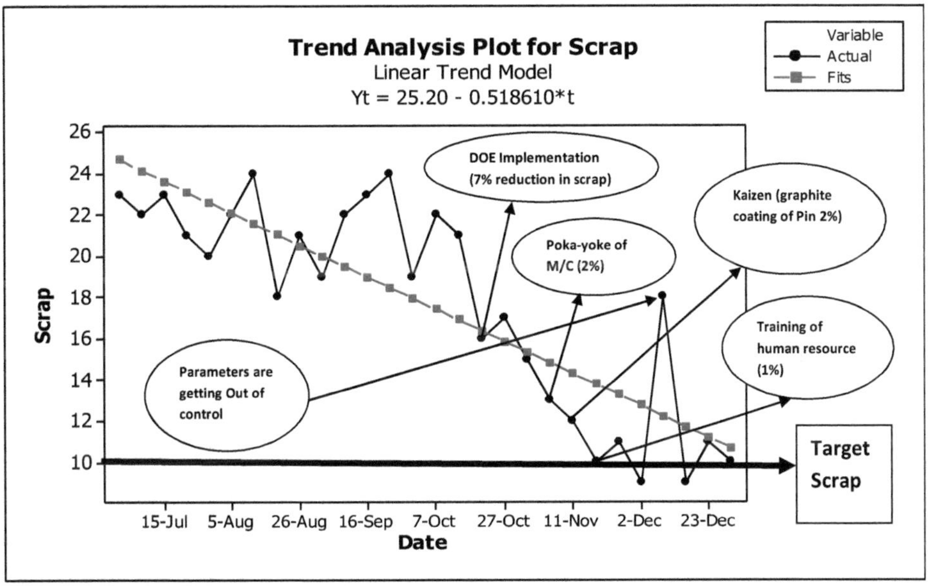

Trend Analysis Plot for Scrap
Linear Trend Model
$Y_t = 25.20 - 0.518610*t$

4.7.3 Inferences from I-Phase

Some major inferences from this phase are:

- More than 58.3% scrap reduction has been achieved through DoE approach alone which is a structured and organized method for determining the relationships among factors affecting a process and its output.

- It has been realized that DoE can offer return that are four to eight times greater than the cost of running the experiments in one-factor-at-a-time pattern.
- It is quite rare to see improvement in quality and productivity of process parameters through Design of Experiments, specifically in Indian foundries.
- The industrial engineering tools like Kaizen and Poka-yoke have also been effective along with DoE, to further improve and fool proof the existing foundry process.
- DoE has helped to investigate the best combinations of process parameters, changing quantities and combinations in order to obtain statically reliable results.
- A systematic route can be followed to find solutions to industrial problems with greater objectivity by means of experimental and statistical techniques.
- In total, a net 12% reduction in scrap was achieved through implementation of proposed improve phase framework.

4.8 Control Phase

The **C** in DMAIC is about controlling the 'vital few' variables, typically the top two or three, that have been identified in the Analyze and Improve phases of the project (Yang and Yeh, 2007). The primary objective of this phase is to ensure the gains and maintain the benefits for a long period. It is necessary to standardize and document procedures, train employees and communicate the project's results. By the time the project reaches the control stage, it would have reached the end of input filtering or funneling process (De Feo, 2000). After all, one of the main reasons for defects is variation and if inputs are not in control then there is a chance that their settings may change and cause variation in process. The steps needed to ensure control of the inputs can be outlined in a control plan. This will help to see which inputs require the maximum attention and the most stringent controls. Another big task that is performed during the control phase is the monitoring of improved process. There may be some improvements that have not worked, so there is a need to implement them in a better way. This phase unfortunately is one of the most overlooked phases in the methodology. After the Improve Phase not much attention is paid to the Control Phase, which leads to short-lived improvements. This is probably the last part of the DMAIC process. If the improvements which should have been verified during the Improve stage are not controlled then all the work becomes in vain. In addition, this is the opportunity to transfer the knowledge to similar areas and so the benefits flow wider than simply the original project (Mathews and Sears, 2005). Due to these facts almost 31% of the projects have been failed because of selecting wrong and absurd tools/techniques in 'Control phase' itself (Bonilla et al., 2008). So to demystify this confusion and further supporting the right selection criteria for various tools/techniques while implementing process control, following integrated methodology has been proposed and validated successfully by executing a case study in non-ferrous foundry.

4.8.1 Proposed Framework for C-Phase

This phase targets the on-going management of the processes that were modified during the previous phase. Unlike the other Six Sigma phases, the Control phase does not have a completion date rather it ensures that the problems that were fixed stay fixed. There is a basic tendency for the process performers to drift back to their previous method of work, after a period of time. A frame work has been proposed to execute control phase systematically (refer figure 4.52). It has two steps. In first step, control phase has been divided into six objectives and then to achieve these, respective tools have been bifurcated through second step. The stated framework almost covers all tasks of control phase like:

- To maintain the gains obtained during improve phase, long after the project has ended up.
- To standardize and document procedures, make sure all employees are trained and communicate the project's results.
- Validate the measuring system discrepancies by performing necessary calibrations.
- Redefine process capability by calculating Cpk and sigma value through DPMO.
- Monitor and revise control documents like; update the control plan and work instructions etc. Manage risks by forming proper FMEA tables.
- Check the scope of achieved improvements by finding similar areas on production floor for increasing benefits exponentially.
- Review all the direct and indirect financial and non-financial benefits incurred from present study.
- Sustaining the improved process through refined checks and practices.

Figure-4.52 Proposed Framework for Control Phase

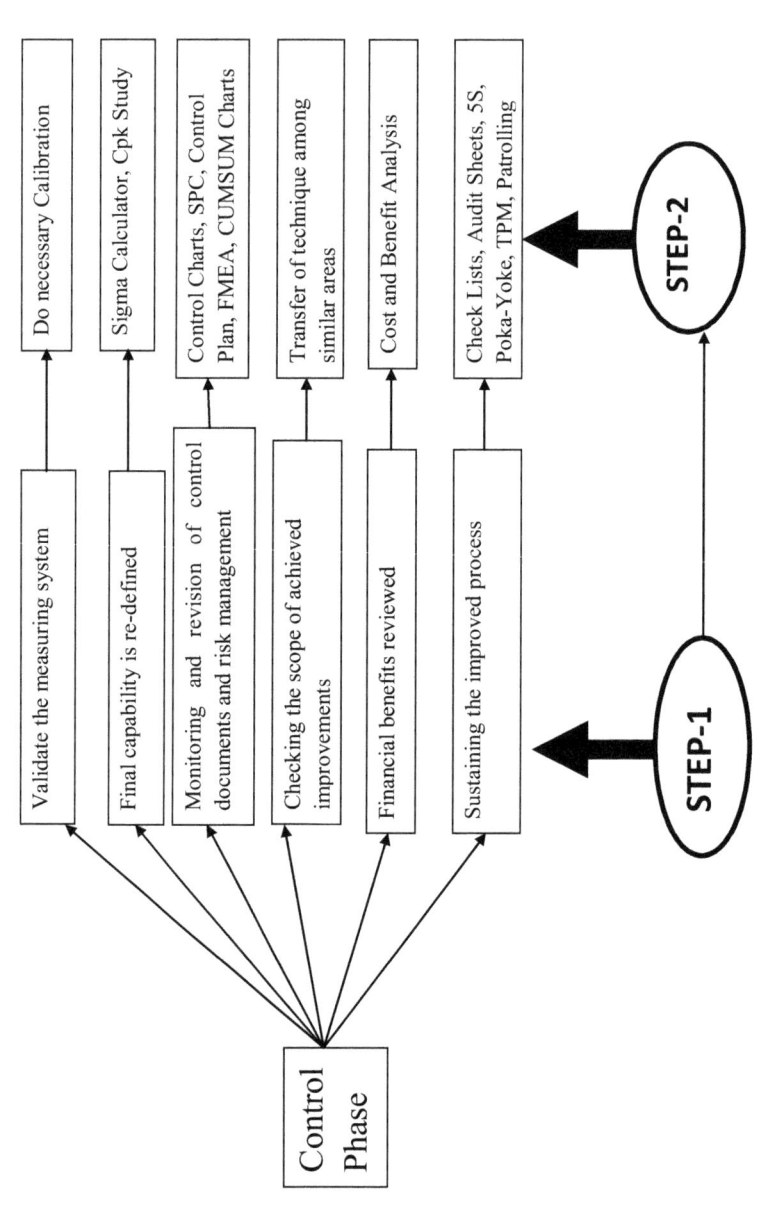

4.8.2 Case Findings (C-Phase)

Tools Used: *Revision of MSA, Six Sigma Calculator, Cpk Calculations, Control Plans, FMEA Table, X Bar R Chart, p-Charts, 5-S, Training Schedule, Process Audit Sheet and Process Indicator Board (refer figure 3.1)*

As per the proposed frame work of this phase, firstly critical measuring equipments of foundry were checked closely to verify the various elements of MSA like bias, linearity, gauge R&R, stability etc. In present study; Immersion Pyrometer, Vac Density Tester and BT Gauge were the major measurement instruments that needed to be calibrated on time. Their accuracy and precision must be ensured for future. A 'MSA Schedule Chart' was drafted (see Annexure 6) to monitor and fulfill the need of calibration with time. Next task was to realize the improved sigma level of given process. For this, one month scrap data was captured after running improved foundry process for full month. Depending upon the DPMO sigma calculator was again used to calculate the sigma level of improved process (refer figure 4.53).

Figure-4.53 Improved Sigma Value

Production Results of Dec-2010	
Total number of Machined parts	14079
Scrap type	**Nos.**
Bottom Thickness Defect	410
Blow Holes	320
Cold Lap	95
Depression	85
Hydrogen Porosity	40
Shrinkage	350
Defective Pin Hole	166
Total Scrap in December	1466
Nos. of Opportunities	7
DPMO	14875
Sigma Level of Process	**3.67**
Yield (%)	96.96

Sigma level improved by 0.24 in first attempt and it is significant for a piston foundry. The yield of foundry also increased from 76.9 to 96.9 percent, as less scrap creates opportunity to cast more good pistons per shift. Figure 4.54 shows improvements achieved in process capability index.

Figure-4.54 Process Capability of Improved Process

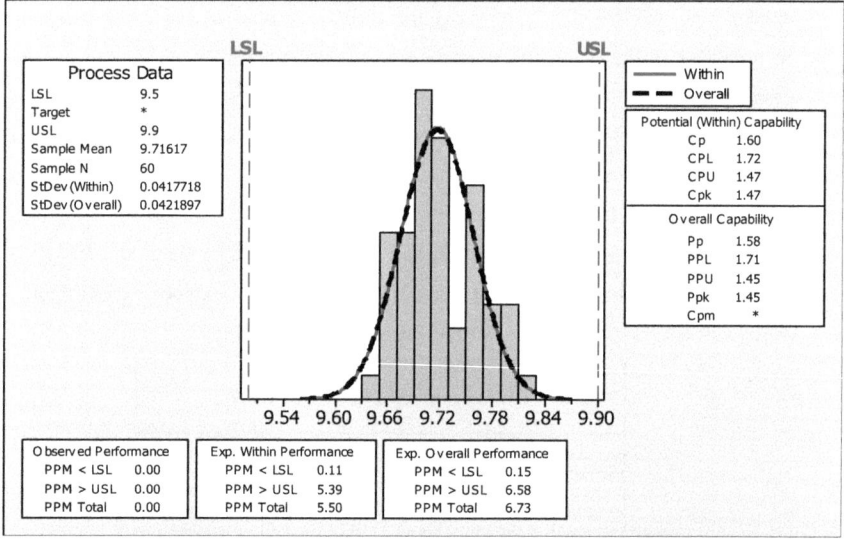

The value of Cpk has been raised to 1.47 that indicates more repetitive and accurate nature of process. The Annexure 8 shows the repeatability and reproducibility of modified casting process of H-273 pistons with respect to BT variation.

In order to control the concerned parameters at their improved values, necessary modification in 'Control Plan' was done. After brain storming maximum tolerance limits for metal temperature, water temperature and cooling time were fixed. From response optimizer it was inferred that die design needed a change to increase in runner-riser volume from 255cm^3 to 281.7 cm^3 to reduce scrap. For this, a separate local riser of volume 26.7cm^3 was added above the top of piston casting to avoid frequent blowholes and shrinkage defects on top and ring zone of piston castings. Extra runner and riser volume on top, helped to provide proper back-feeding of molten metal while solidification of top and ring zone portion of casting and hence helped to avoid around 25% of casting defects (refer figure 4.55). It was recommended to check 5-S condition of casting work station daily by visual inspection. Rules for ladle alignment and ladle height at the time of pouring had been quoted in control plan. After every two hours, reading of molten temperature was taken with the help of pyrometer and control over pouring speed was suggested. DRO was installed with sensor to monitor the cooling water temperature on continuous basis. FMEA table had been generated for calculating risk priority number (RPN) of each foundry operation (refer annexure 7).

Figure-4.55 Casting with Increased Volume of R&R

 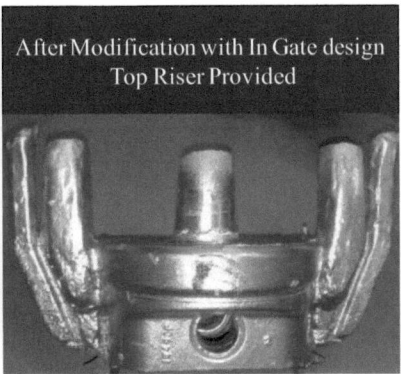

FMEA tables not only categories the criticality of process parameters but also suggests corrective and preventive actions needed at the time of failure of any process parameter. Further to monitor the bottom thickness variation (as a major variable) X-bar and R chart had been drawn. For this a continuous data at the interval of 2 hours was collected (refer table 4.17). The SQC operator was appointed on the casting workstation and was made responsible to check the casting scrap due to BT variation, on shift basis.

Table-4.17 Data of BT Variation

Bottom thickness variation of H-273 Pistons (Dec-2010)

	Mean-9.70mm						Upper limit- 9.95mm			Lower limit-9.48mm		
Time	7:00 AM	9:00 AM	11:00 AM	1:00 PM	3:00 PM	5:00 PM	7:00 PM	9:00 PM	11:00 PM	1:00 AM	3:00 AM	5:00 AM
X1	9.72	9.82	9.77	9.55	9.43	9.29	9.50	9.75	9.74	9.31	9.94	9.94
X2	9.85	9.90	9.79	9.50	9.35	9.30	9.54	9.82	9.94	9.57	9.31	9.31
X3	9.90	9.80	9.65	9.45	9.30	9.31	9.65	9.94	9.31	9.74	9.57	9.48
X4	9.65	9.50	9.56	9.41	9.31	9.39	9.74	9.55	9.57	9.31	9.74	9.57
X5	9.55	9.90	9.70	9.38	9.28	9.40	9.72	9.57	9.74	9.94	9.49	9.74
Mean	9.73	9.78	9.69	9.46	9.33	9.34	9.63	9.72	9.66	9.57	9.61	9.60
Range	0.35	0.4	0.23	0.17	0.15	0.11	0.24	0.39	0.63	0.63	0.63	0.63

The corresponding graph with upper and lower limits was generated as described in figure 4.56. This graph would monitor the special variation in BT regularly and help to reduce the scrap due to over BT.

Figure 4.56 Control Chart for BT Defect

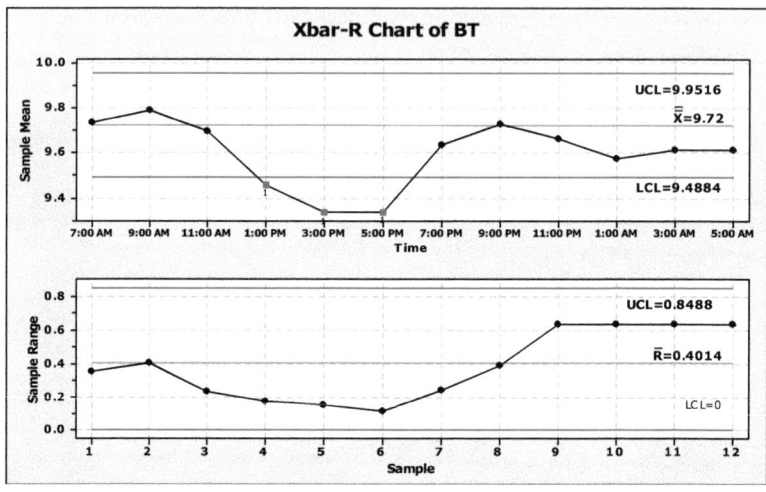

To monitor overall casting scrap due to all other reasons, p-chart had been drawn daily by inspection supervisor. The one month data for p charts was collected (refer annexure 9) and graph had been drawn (see figure 4.57).

Figure 4.57 p-Chart for Monitoring Casting Scrap

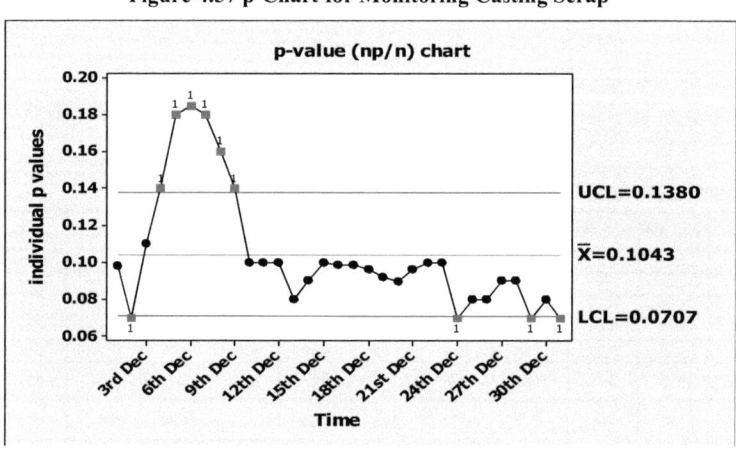

It predicts the day to day status of scrap in piston foundry. It shows after improvement, mean scrap is approximately 10.4% and it gives indication/alarm when the process scrap goes beyond 13.8%. These control measures had successfully run the piston foundry for consecutive two months at around 10% scrap only, leading to savings of around 8.78 lakhs per annum. It was seen that other H family pistons like H-519 and H-749 were quite similar to H-273 pistons, as far as their designs are concerned due to which their respective dies and process parameters had lot of similarities. The same DMAIC methodology was implemented for H-749 and H-519 pistons, in next two months. It resulted in additional financial benefits of Rs 13.1 lakhs and Rs 8.7 lakhs for H-519 and H-749 pistons, respectively. Collectively this amount was even more than the initial savings achieved from H-273 pistons (refer annexure 10). As foundries are energy intensive units, so energy savings produced from above Six Sigma implementations can also be assessed as indirect savings achieved.

At the end of control phase, some sustaining measures were brought in day to day working practice, for long term benefits. This part also included some 'Value Engineering' proposals for foundry operations, like:

- Proper 'On the Job Training Schedule' was made to make aware all the related shop floor staff, so that their code of conduct became positive and responsive to tackle DMAIC responsibilities.
- For more knowledge and to understand the significance of Six Sigma improvements a 'Six Sigma Corner' was developed in the middle of production floor and was operated by a black belt champion.
- A comprehensive check list was designed to cover every aspect of each phase and helpful to take vital steps in right time.
- Process Audit Sheet was formulated (refer figure 4.58).
- A 'Casting Process-Indicator Board' was specifically designed to execution all the casting activities of pistons in a desired sequence (refer figure 4.59). It also supported vital process variables like die temperature, cooling water temperature and cycle time with series of successive indicators that had been lighted up, once the relevant check or activity was performed or ensured successfully. Proper hooters had also been installed with the board to warn the machine operators in case of missing of any casting activity or its parameter settings. It really made the whole die casting process,

mistake proof and insured optimization of relevant variables efficiently and as per process requirements.

Figure-4.58 Process Audit Sheet

Parameter	Specification	Observation	Remarks
Scrapping of crucible walls	Clean walls	ok	was done before tapping of metal in HF
Drossing off with Cov-11	Measuring Spoon	Visual inspection	YES 329 gms
Holding Furnace Clean from S	visual clean	Not Clean	Not ok
Degassing - - Cl2	10 ± 2 minutes	10min 11 sec	OK
N₂ / Argon	Time = 6 ±1 min.		
	Time = 12 ±2 min.	12 minute	Done with MDU (6 Min Auto Cycle),Done Twice
	Pressure=2-4 bar	2 KG/CM2	OK
	Flow Rate=6±2 LPM	6 LPM	OK
Waiting time	5 Mins minimum	3 Min 50 sec	Not ok
Sample probe	Visual	Sample Piston turned	Gas free, ok
Density Index	upto 1.0	0.6	Ok
Casting temp.	880±10°C	HF 3 Line 2-870°C	Ok
Die cooling cycle	as per Spec.	ok	Ok
Luff Sticks/air vents Open	open	Luff stick not present in left tool	left tool opened and sent for providing luff stick
Tool wear out	No wear	NO WEAR	OK
Casting Spoon	Refer specification	Coated	OK
Die Paint/ Padding	Hatching Lines not Visible	NOT OK	CASTER called and advised to refill the padding
Core grouping	Proper steps	No Tool Shifting	OK

Figure-4.59 Process Indicator Board

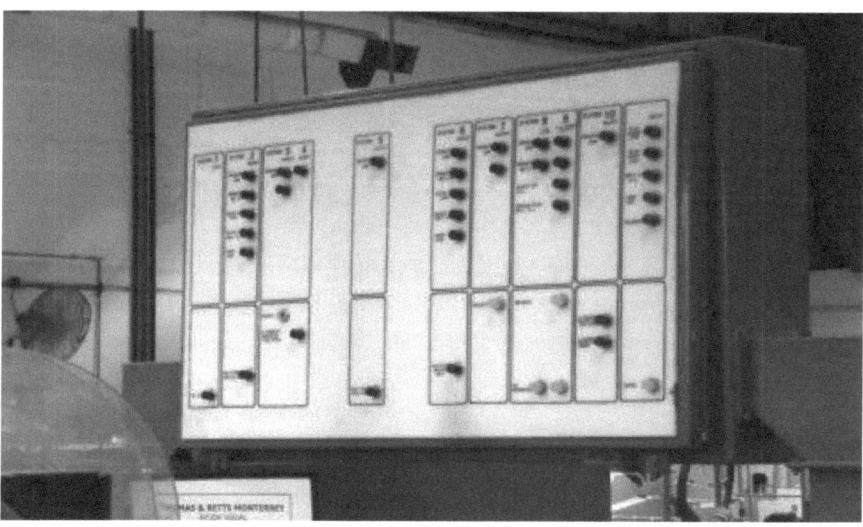

- Three patrol teams of casting operators were constituted to vigil the critical foundry processes and their vital parameters, while patrolling in each shift.
- To sustain the overall improvements acquired by DMAIC process, one executive level person was deputed to visit shop floor once a week, so that value adding activities were performed in a more controlled and consistent manner.

4.8.3 Inferences from C-Phase

Some major inferences from this phase are:

- MSA schedule chart had been chalked out for monitoring the calibration and repair record of measuring instruments.
- Sigma level after improve phase had been raised to 3.67 as calculated from Sigma calculator.
- Cpk with respect to BT variation became 1.47 and low spread of bell shape curve had indicated more controlled casting process after improve phase.
- BT variation would be checked and tracked in future through shift wise X bar & Range chart.
- p-chart would monitor the overall casting scrap month wise. Control plan was updated with optimum values of casting parameters. Necessary changes had been incorporated in die In-gate design to rise the net volume of runner and riser, strategically.
- Process indicator board had fool proofed the whole casting process through installation of vital audio-visual indications and checks.
- Daily process audit sheet had further sustained the foundry operations and ensured the high degree of control on shop floor.

4.9 Result Appraisal

Following are some major results achieved during implementation of sequential Six Sigma phases through proposed frameworks;

- The existing high scrap (almost 22%) was defined as most serious issue as far as overall productivity of pistons is concerned.
- Casting scrap had been found primarily due to defects like; shrinkage, bottom thickness variation, blow holes, porosity, pin hole defect and cold lap.

- Process parameters (SSVs) like; in-gate design, dimensional inaccuracy of machine, die temperature, die coating thickness, discharge of cooling water, metal sticking on pins, alloy temperature, shift dependency and operator skill were measured as main reasons of above defects in piston castings.
- Although during Measurement System Analysis (MSA), gauge R&R of BT gauge had been found ok, yet Immersion Pyrometer (Temp. measuring device) was biased and Vac-tester (Metal density checker) was coming out to be un-stable. So both the measuring equipments were repaired and calibrated suitably before starting analyse phase.
- After analysis of all the eleven SSVs with appropriate statistical tools, seven critical process parameters emerged as responsible for high scrap. These were; in-gate design (volume of runner & riser), operator skill, discharge of cooling water, die temperature, delay time, metal sticking on pins and dimensional in accuracy of machine.
- In Improvement Phase, four casting process parameters (i.e. volume of R&R, die temperature, discharge of cooling water and delay with in process) had been optimized to bring a significant reduction of 7% in scrap.
- The machine's dimensional in-accuracy was enhanced by altering the design of machine through Poka-Yoke principles and this helped to achieve a net reduction of around 2% in scrap.
- The metal sticking on pins was avoided by introducing a standard process of pin cleaning, prepared by using kaizen rules of continues improvement and hence brought further reduction of 2% in scrap.
- Then a proposed training schedule reduced the scrap by around 1%. So overall 12% (approximately) reduction in scrap had been achieved during improvement phase of this study.
- Earlier existing sigma level was 3.43 and after improvement it had been raised by 0.24 only, but even this has resulted in savings of around 8.78 lakhs per annum for H-273 pistons.
- Overall scrap of foundry had been monitored by drawing shift wise p-chart and controlled up to 10.43% only. BT variation was also monitored at casting machine

itself, by chalking out X bar & Range chart hourly. Cpk study had also been conducted by SQC operator after every 4 shifts, for time to time process capability-checking.

- At the end whole casting process was controlled by suggested ways at different steps. Profit had been exponentially raised by transferring the similar improvements to other similar part numbers. Net overall profit harvested was Rs. 30.7 lakhs per annum by implementing DMAIC approach as per proposed frameworks for each phase (refer Annexure 10).

Exercise

1. What do you mean by Piston? Where it is used?
2. Elaborate DMAIC approach of Six Sigma? Discuss various tools used in its each phase?
3. What is the importance of Define Phase in any Six Sigma Project?
4. Explain various pit falls which must be taken care of while executing Measuring Phase?
5. Classify different Analytical tools in Quantitative or Qualitative categories?
6. Write a step wise procedure to execute DoE in Indian foundries?
7. What do you mean by Why-Why Analysis? Discuss its applications?
8. Elaborate Poka-Yoke? How it can be taken as an improve phase tool?
9. Illustrate the Procedure to implement Define Phase & Measure Phase in Foundries?
10. What do you mean by Pin Defect (In piston castings)? How one can check it effectively?
11. Give step wise procedure to conduct the Control Phase in Indian Foundries?
12. What is the importance or need of executing Control Phase while Six Sigma projects?

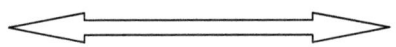

CHAPTER - 5   IMPLICATIONS & SCOPE

5.1 Conclusions

Since its introduction in the 1990's, Six Sigma has become the buzzword in both the manufacturing and service industries. The various methodologies used in Six Sigma are based on a disciplined and data driven approach that help in eliminating defects and achieving near perfection by restricting the number of possible defects to less than 3.4 defects per million. Although many companies have been successful in reducing the number of defects through Six Sigma projects, still myths against the efficacy of Six Sigma in all aspects of business processes do not seem to die down. The full potential of Six Sigma has not been realized so far because many competent SMEs have still not implemented Six Sigma programs. These enterprises have all the resources to implement such programs, but are often wary of initiating Six Sigma as they believe that it is meant only for large organizations. To dispel this myth, a case study on a medium scale foundry unit was planned for implementing Six Sigma as a project-driven strategy to improve productivity levels by reducing scrap.

A comprehensive literature survey was carried out on various aspects of Six Sigma like; origin, development, themes, tools and status in Indian and foreign companies. It was concluded that big companies and corporations have been the primary beneficiaries, though small and medium sectors are getting attracted towards it. Manufacturing sector has made more extensive use of Six Sigma as compared to process and service sectors. In India, like in other countries, Six Sigma has been implemented extensively but strangely no foundry industry has adopted this strategy despite the presence of high scrap/rejection rate and low productivity levels. All these research gaps further motivated to carry out present research in medium scale foundry unit. During the case study, after analysis and deliberations, some practical difficulties were anticipated to be responsible for Six Sigma failures and these are:

- Wrong project selection
- Data collected is awful
- Ignoring MSA
- Scarcity of training in Six Sigma

- Wrong selection of tools & techniques while executing various phases of DMAIC methodology.
- Project never finished.
- Project finished but improvement/saving was insignificant.
- Project finished but solution not implemented.
- No long term implementation of solution.
- Lack of management support.
- Need of cultural changes for sustaining benefits long term.

So a comprehensive research methodology was proposed for successful implementation of Six Sigma in Indian SMEs, without ignoring the constraints. After a successful implementation of project based Six Sigma approach, certain Six Sigma imperatives were identified for the Indian foundries and these are:

- This book validates the 'Project Based' approach for Six Sigma implementation in Indian context, after implementing DMAIC in a piston foundry. Adopted approach encourages team working among multi-disciplinary members to collaborate and create opportunities for individuals and organizations to grow and prosper. The benefits incurred from these projects will automatically motivate the management and work force for their mutual betterment and seek more similar projects in future.
- Execution of each phase seems to be more important than various tools or techniques used under them. There must be a strategy to conduct all the five phases of DMAIC approach, depending upon the given conditions and resources of the system. The proposed frame works for every phase helps to select the right tool or technique as per the real requirement and thus reduces the risk of failure of Six Sigma projects due to wrong selection of tools.
- Right selection and use of statistical and non-statistical tools during Six Sigma has raised the issue of 'on-job training on Six Sigma'. The challenge for all organizations is to integrate Six Sigma into their core business processes and make it as their culture. Basic tenets for Six Sigma should be:
  - Constitute a multi-functional group that will be responsible for Six Sigma documentation.
  - Teach Six Sigma by example.

- Commission the group to produce reports or documents on process flow diagrams, product performance, appearance, finished products and customer requirements etc.
- Define phase needs management and also involves many statistical tools/ techniques. So there is always a risk of choosing wrong tools due to ignorance. This can lead to failure of this approach and it may involve lot of monetary losses. The proposed frame work tries to simplify the D phase of Six Sigma and categories the tools/techniques with respect to their utility. The framework has further been successfully validated for its effectiveness through the case study.
- 'Operation Measurement' has now emerged as a discipline in itself and so present work has framed a rigid road map to guide and monitor the accurate selection of measurement tools by proposed operation measurement methodology. Proper execution and interpretation of each step has also been done to reduce the time and effort by proper management of measurement tools and techniques.
- Industrial analysis has now emerged as a discipline in itself and is motivated by its own programme of development. So a roadmap has been chalked out to guide and monitor the accurate selection of analytical tools. Their proper execution and interpretation has also been discussed through the case study.
- Presently, in developing countries the productivity index is quite low in foundries and this study explores the significance of DoE for SMEs. The proposed methodology gives a framework to reduce the time and effort to implement accurate DoE by having appropriate knowledge of analytical tools/techniques used in a more applicable form.
- This book has also proposed a more applicable methodology to execute operation controls comprehensively. This can act as a road map for short listing the relevant control tools especially for Six Sigma learners, people working in industry and professional-statisticians.
- There is nothing as bad as operating a process whose basics are not clear. Thus it is essential that every person in the organization understands customer requirements and has a general idea of how the process will assist in satisfying the customer.
- Poor communication among concerned work force working on projects may result in one group accusing the other of handing over one job or another to them, or completely avoiding responsibility. In constituting the team, it is essential to identify employees with

strong communication skills. In unionized environments, union members who hold publicity responsibilities may be useful for this assignment. Communication should also be improved within and beyond the company. The communication group should seek to disseminate information on the ongoing Six Sigma project to vendors and improve company relationships with it.

- Top management commitment is a key success factor in the implementation of Six Sigma. Such commitment manifests in top management's regular attendance at meetings, encouraging team members to be committed to work and emphasizing the need for goal achievement in employees' tight schedules.
- Individuals' knowledge must be fully utilized for high quality decision-making. In implementing Six Sigma projects, the advantage of expert opinion within the organization must be used. Naturally, there may be some team members willing to provide tips and information in solving Six Sigma problems. Well-planned and implemented training courses, handled firstly by outside experts and then by in-house experts, should be carried out for the various categories of employees. Courses designed for beginners, intermediate and advanced employees in Six Sigma knowledge are essential.

This book has dwelled on the use of Six Sigma within Indian foundry SMEs. It provides a model to effectively guide the implementation of Six Sigma programs to reduce variation or waste from the foundry operations. This work would inspire Indian foundry SMEs to fetch benefit from this strategy in a 'Project Based' approach rather then planning, training or investing here and there for accumulating various resources required for Six Sigma implementation. The pressure to pursue new ways of thinking as a source of competitive advantage is rising day by day. More research in this area is still necessary to contribute to the science and practice of implementation of Six Sigma or any other process improvement model, to reduce waste and create value, particularly in developing nations like 'India'.

5.2 Scope

Six Sigma, a systematic approach for quality improvement and business excellence, has been widely publicized in recent years as the most effective means to combat quality problems and win customer satisfaction, but still it is in primary stage as far as Indian industries are related and in this context the following aspects need attention in future;

- Beside non-ferrous foundries, Six Sigma approach can be explored for ferrous foundries to bring breakthrough in rejections and increase yield per annum.
- Six Sigma can also be used in energy intensive foundries, as it not only enhances productivity by process improvement but also it is a step to create 'zero defect foundries' which indirectly reap huge energy / power savings.
- It can be implemented in green sand foundries specifically in sand re-use plants for not only to reduce the lead time of process but also to cast cost effective castings.
- Apart from foundry industries, other manufacturing sectors like forging, forming, welding and machining industries can also take benefit to lean their business operations.
- Six Sigma should further be explored in service sector like hospitals, education institutes, banking, traffic etc in India.
- Through an extensive literature search, it was observed that very little documentation exists in the application of Six Sigma to education sector. In one case, it was limited to statistics education. Clearly, Six Sigma researchers have many questions to answer.

Exercise

1. Talk about different Myths prevailing about Six Sigma in Indian scenario?
2. Discuss how this book is helpful to demystify the existing myths among Indian foundries?
3. Write at least five major conclusions made by this book?
4. Discuss the role of top management as far as Six Sigma implementation is concerned?
5. What is the importance of Training in Six Sigma? Explain with respect to foundries?
6. Write a short note on 'Effective Project Selection'?
7. Tell briefly about the success of case study discussed in this book?
8. Give your opinions about naval procedures presented to execute different DMAIC phases?
9. Write at least three scopes of this book in near future?
10. Give limitations of present work?

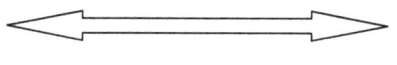

BIBLOGRAPHY

1. Aboelmaged, M.G. (2009), "Six Sigma quality: a structured review and implications for future research", International Journal of Quality and Reliability Management, Vol. 27, No. 3, pp. 268-317.
2. Agarwal, R. and Bajaj, N. (2008), "Managing outsourcing process: applying Six Sigma", Business Process Management Journal, Vol. 14, No. 6, pp. 829-37.
3. Aggogeri, F. and Gentili, E. (2008), "Six Sigma methodology: an effective tool for quality management", International Journal of Manufacturing Technology and Management, Vol. 14, Nos 3/4, pp. 289-98.
4. Al-Aomar, R. (2006), "A simulation-based DFSS for a lean service system", International Journal of Product Development, Vol. 3, Nos 3/4, pp. 349-68.
5. Al-Mishari, S. and Suliman, S. (2008), "Integrating Six-Sigma with other reliability improvement methods in equipment reliability and maintenance applications", Journal of Quality in Maintenance Engineering, Vol. 14, No. 1, pp. 59-70.
6. Amer, Y., Ashraf, M., Luong, L., Lee, S. and Wang, W. (2007), "A systems approach to order fulfilment using design for Six Sigma methodology", International Journal of Business and Systems Research, Vol. 1, No. 3, pp. 302-10.
7. Andersson, R., Eriksson, H. and Torstensson, H. (2006), "Similarities and differences between TQM, Six Sigma and lean", The TQM Magazine, Vol. 18, No. 3, pp. 282-96.
8. Andre A de Waal, Robert Goedegebuure, Patricia Geradts, (2011) "The impact of performance management on the results of a non-profit organization", International Journal of Productivity and Performance Management, Vol. 60, No. 8, pp. 89-99.
9. Antony, J. (2002), "Design for Six Sigma: a breakthrough business improvement strategy for achieving competitive advantage", Work Study, Vol. 51, No. 1, pp. 6-8.
10. Antony, J. (2004a), "Six Sigma in the UK service organizations: results from a pilot survey", Managerial Auditing Journal, Vol. 19, No. 8, pp. 1006-13.

11. Antony, J. (2004b), "Some pros and cons of Six Sigma: an academic perspective", The TQM Magazine, Vol. 16, No. 4, pp. 303-06.
12. Antony, J. (2006), "Six Sigma for service processes", Business Process Management Journal, Vol. 12, No. 2, pp. 234-48.
13. Antony, J. (2008), "Can Six Sigma be effectively implemented in SMEs?", International Journal of Productivity and Performance Management, Vol. 57, No. 5, pp. 420-23.
14. Antony, J. and Banuelas, R. (2002), "Key ingredients for the effective implementation of Six Sigma program", Measuring Business Excellence, Vol. 6, No. 4, pp. 20-27.
15. Antony, J. and Desai, D. A. (2009), "Assessing the status of six sigma implementation in the Indian industry: Results from an exploratory empirical study", Management Research News, Vol. 32, No. 5, pp. 413-23.
16. Antony, J. and Kaye, M. (1996), "Optimisation of core tube life using experimental design methodology", Journal of Quality World, pp. 42-50.
17. Antony, J., Antony, F. and Kumar, M. (2007), "Six Sigma in service organizations: benefits, challenges and difficulties, common myths, empirical observations and success factors", International Journal of Quality & Reliability Management, Vol. 24, No. 3, pp. 294-311.
18. Azadeh, A., Sadegh, M. and Sangari, M.S. (2010) 'A metaheuristic method for optimizing inspection strategies in serial multistage processes', Int. J. Productivity and Quality Management, Vol. 6, No. 3, pp.289–303.
19. Bakshi, Y., Singh, B.J., Singh, S.S. and Singla, R. (2012), "Performance Optimization of Backup Power Systems through Six Sigma: A Case Study", IETET-2012, GITA Institutes, Kurukshetra (Held in Nov 2012)
20. Bandyopadhyay, J. and Lichtman, R. (2007), "Six Sigma approach to quality and productivity improvement in an institution for higher education in the United States", International Journal of Management, Vol. 24, No. 4, pp. 802-08.
21. Banuelas, R., Antony, J. and Brace, M. (2005), "An application of Six Sigma to reduce waste", Quality and Reliability Engineering International, Vol. 21, pp. 553-70.

22. Basu, R. (2004), "Six Sigma to operational excellence: role of tools and techniques", International Journal of Six Sigma and Competitive Advantage, Vol. 1, No. 1, pp. 44-64.
23. Bayle, P., Farrington, M., Sharp, B., Hild, C. and Sanders, D. (2001), "Illustration of Six Sigma assistance on a design project", Quality Engineering, Vol. 13, No. 3, pp. 341-48.
24. Behara, R.S., Fontenot, G.F. and Gresham, A. (1995), "Customer satisfaction measurement and analysis using Six Sigma", International Journal of Quality & Reliability Management, Vol. 12, No. 3, pp. 9-18.
25. Bendell, T. (2006), "A review and comparison of Six Sigma and the lean organizations", The TQM Magazine, Vol. 18, No. 3, pp. 255-62.
26. Benedetto, A. (2003), "Adapting manufacturing-based Six Sigma methodology to the service environment of a radiology film library", Journal of Healthcare Management, Vol. 48, pp. 263-80.
27. Bhardwaj, A., Gupta, A. and Kanda, A. (2010a), "Fundamental Concepts of Theory of Constraints: An Emerging Philosophy", World Academy of Science, Engineering and Technology 70, pp. 687-693.
28. Bhardwaj, A., Singh, L.P. and Deepak, K.K. (2010b), "Occupational exposure in small and medium scale industry ", International Journal of Noise and Health", Vol. 12, No.46, pp. 37-48.
29. Bhasin, S. (2011) "Measuring the Leanness of an organisation", International Journal of Lean Six Sigma, Vol. 2, No. 1, pp.55 – 74
30. Black, K. and McGlashan, R. (2006), "Essential characteristics of Six Sigma black belt candidates: a study of US companies", International Journal of Six Sigma and Competitive Advantage, Vol. 2, No. 3, pp. 301-12.
31. Black, K. and Revere, L. (2006), "Six Sigma arises from the ashes of TQM with a twist", International Journal of Health Care Quality Assurance, Vol. 19, Nos 2/3, pp. 259-66.
32. Bonilla, C., Pawlicki, T., Perry, L. and Wesselink, B. (2008), "Radiation oncology: Lean Six Sigma project selection based on patient and staff input into a

modified quality function deployment", International Journal of Six Sigma and Competitive Advantage, Vol. 4, No. 3, pp. 196-208.

33. Box, T. (2006), "Six Sigma quality: experiential learning", S.A.M. Advanced Management Journal, Vol. 71, No. 1, pp. 20-23.

34. Brett, C. and Queen, P. (2005), "Streamlining enterprise records management with lean Six Sigma", Information Management Journal, Vol. 39, No. 6, pp. 52-62.

35. Brewer, P. (2004), "Six Sigma helps a company create a culture of accountability", Journal of Organizational Excellence, Vol. 23, No. 3, pp. 45-59.

36. Byrne, G. (2003), "Ensuring optimal success with Six Sigma implementations", Journal of Organizational Excellence, Vol. 22, No. 2, pp. 43-50.

37. Camgoz-Akdag, H. (2007), "Total quality management through Six Sigma benchmarking: a case study", Benchmarking: An International Journal, Vol. 14, No. 2, pp. 186-201.

38. Carnell, M. and Lambert, J. (2000), "Organisational excellence through Six Sigma discipline", Measuring Business Excellence, Vol. 4, No. 2, pp. 18-25.

39. Caulcutt, R. (2001), "Why is Six Sigma so successful?", Journal of Applied Statistics, Vol. 28, Nos 3/4, pp. 301-06.

40. Chakrabarty, A. and Tan, K. (2007), "The current state of Six Sigma application in services", Managing Service Quality, Vol. 17, No. 2, pp. 194-208.

41. Chan, W., Chiu, W-T., Chen, W-M., Lin, M.-F. and Chu, B. (2005), "Applying Six Sigma methodology to maximise magnetic resonance imaging capacity in a hospital", International Journal of Healthcare Technology and Management, Vol. 6, No. 3, pp. 321-30.

42. Chang, C.-M. and Su, C.-T. (2007), "Service process design and/or redesign by fusing the powers of design for Six Sigma and lean", International Journal of Six Sigma and Competitive Advantage, Vol. 3, No. 2, pp. 171-91

43. Chang, K. and Wang, F. (2008), "Applying Six Sigma methodology to collaborative forecasting", International Journal of Advanced Manufacturing Technology, Vol. 39, Nos 9/10, pp. 1033-44

44. Chatterjee, A. (2003), "Innovating growth through 'Six Sigma': a strategic approach for combining robustness with flexibility", Global Journal of Flexible Systems Management, Vol. 4, No. 3, pp. 33-37
45. Chaudhery, P.K. (2010) 'Annual supplement foreign trade policy 2009–2014 – major highlights', Indian Foundry Journal, August, Vol. 56, No. 11, pp.95–97.
46. Chen, K., Lin, J. and Chang, C. (2005), "Taiwan: improving radiography through application of Six Sigma techniques", Journal for Healthcare Quality, Vol. 27, No. 3, pp. 44-52.
47. Choudhary, D. and Jain, P.K. (2010) 'Short-term decision making in foundry unit: a case study', Indian Foundry Journal, Vol. 56, No. 11, pp.34–39.
48. Coronado, R. and Antony, J. (2002), "Critical success factors for the successful implementation of Six Sigma projects in organisations", The TQM Magazine, Vol. 14, No. 2, pp. 92-99.
49. Craven, E., Clark, J., Cramer, M., Corwin, S. and Cooper, M. (2006), "NewYork-Presbyterian hospital uses Six Sigma to build a culture of quality and innovation", Journal of Organizational Excellence, Vol. 25, No. 4, pp. 11-19.
50. Dahlgaard, J. and Dahlgaard-Park, S. (2006), "Lean production, Six Sigma quality, TQM and company culture", The TQM Magazine, Vol. 18, No. 3, pp. 263-81.
51. Daniel F., (2011) "Statistical, technical and sociological dimensions of design of experiments", The TQM Journal, Vol. 23, No. 4, pp.435 – 445.
52. Das, N., Gauri, S. and Das, P. (2006), "Six Sigma principles in marketing: an application", International Journal of Six Sigma and Competitive Advantage, Vol. 2, No. 3, pp. 243-62.
53. Dasgupta, T. (2003), "Using the Six-Sigma metric to measure and improve the performance of a supply chain", Total Quality Management and Business Excellence, Vol. 14, No. 3, pp. 355-66.
54. Davison, L. and Al-Shaghana, K. (2007), "The link between six sigma and quality culture: an empirical study", Total Quality Management and Business Excellence, Vol. 18, No. 3, pp. 249-65.

55. De Feo, J. (2000), "Six Sigma: new opportunities for HR, new career growth for employees", Employment Relations Today, Vol. 27, No. 2, pp. 1-6.
56. De Feo, J. and Bar-El, Z. (2002), "Creating strategic change more efficiently with a new design for Six Sigma process", Journal of Change Management, Vol. 3, No. 1, pp. 60-80.
57. De Koning, H. and De Mast, J. (2006), "A rational reconstruction of Six-Sigma's breakthrough cookbook", International Journal of Quality & Reliability Management, Vol. 23, No. 7, pp. 766-87.
58. Dedhia, N.S. (2005), "Six Sigma basics", Total Quality Management and Business Excellence, Vol. 16, No. 5, pp. 567-74.
59. Desai, D. (2006), "Improving customer delivery commitments the Six Sigma way: case study of an Indian small-scale industry", International Journal of Six Sigma and Competitive Advantage, Vol. 2, No. 1, pp. 23-47.
60. Does, R., Van Den Heuvel, J., De Mast, J. and Bisgaard, S. (2002), "Comparing non-manufacturing with traditional applications of Six Sigma", Quality Engineering, Vol. 15, No. 1, pp. 177-82.
61. Douglas, P.C. and Erwin, J. (2000), "Six Sigma's focus on total customer satisfaction", Journal for Quality and Participation, Vol. 23, No. 2, pp. 45-49.
62. Drenckpohl, D., Bowers, L. and Cooper, H. (2007), "Use of the Six Sigma methodology to reduce incidence of breast milk administration errors in the NICU", Neonatal Network, Vol. 26, No. 3, pp. 161-66.
63. Dua, A., Batra, U. and Singh, B.J. (2013), "Kinetics of Magnesium Doped Hydroxyapatite: A Review", PTU Jalandhar, pp. 349-352. (Held in Oct 2013).
64. Dua, A., Singh, B.J. and Malik, S. (2014), "Quantitative Analysis for Effect of Copper on Various Mechanical Properties of Ductile Iron during Austempering", International Journal of Engineering Research and Applications (IJERA) Special Issue, pp. 51-56.
65. Dulluri, S. and Raghavan, N.R.S. (2009) 'Stochastic operational planning model for production and distribution in a hi-tech manufacturing network', Int. J. Operational Research, Vol. 5, No. 2, pp.163–189.

66. Edgeman, R. and Dugan, J. (2008), "Six Sigma: from products to pollution to people", Total Quality Management and Business Excellence, Vol. 19, No. 1, pp. 1-9.
67. Ehie, I. and Sheu, C. (2005), "Integrating Six Sigma and theory of constraints for continuous improvement: a case study", Journal of Manufacturing Technology Management, Vol. 16, No. 5, pp. 542-53.
68. Eisenhower C. (2011), "Etienne Taguchi quality specification categories and the computation of Six-Sigma metrics: analytical and service industry anomalies and their managerial implications", International Journal of Six Sigma and Competitive Advantage, Vol. 6, No.4, pp. 243 - 255.
69. Fazzari, A. and Levitt, K. (2008), "Human resources as a strategic partner: sitting at the table with Six Sigma", Human Resource Development Quarterly, Vol. 19, No. 2, pp. 171-80.
70. Feld, K. and Stone, W. (2002), "Using Six Sigma to change and measure improvement", Performance Improvement, Vol. 41, No. 9, pp. 20-26.
71. Ferng, J. and Price, A. (2005), "An exploration of the synergies between Six Sigma, total quality management, lean construction and sustainable construction", International Journal Six Sigma and Competitive Advantage, Vol. 1, No. 2, pp. 167-87.
72. Flott, L. (2000), "Six-Sigma controversy", Metal Finishing, Vol. 98, No. 12, pp. 43-48.
73. Fowlkes, W.Y. (1995), "Creveling, C.M. Engineering Methods for Robust Design; Addison-Wesley", Reading, MA. USA.
74. Frank, S. (2003), "Applying Six Sigma to revenue and pricing management", Journal of Revenue and Pricing Management, Vol. 2, No. 3, pp. 245-54.
75. Freiesleben, J. (2006), "Communicating Six Sigma's benefits to top management", Measuring Business Excellence, Vol. 10, No. 6, pp. 19-27.
76. Freiesleben, J. (2007), "Can Six Sigma claim to be a generic strategy? Reassessing the competitive implications of quality improvement", International Journal of Six Sigma and Competitive Advantage, Vol. 3, No. 3, pp. 248-65.

77. Frings, G. and Grant, L. (2005), "Who moved my sigma? Effective implementation of the Six Sigma methodology to hospitals", Quality and Reliability Engineering International, Vol. 21, No. 3, pp. 311-28.
78. Furterer, S. and Elshennawy, A. (2005), "Implementation of TQM and lean Six Sigma tools in local government: a framework and a case study", Total Quality Management and Business Excellence, Vol. 16, No. 10, pp. 1179-91.
79. Gack, G. and Robison, K. (2003), "Integrating improvement initiatives: connecting Six Sigma for software, CMMI, personal software process, and team software process", Software Quality Professional, Vol. 5, No. 4, pp. 5-13.
80. Gadallah, M.H. (2009) 'Modelling and synthesis using response surface methodology: a comparative study', Int. J. Experimental Design and Process Optimisation, Vol. 1, Nos. 2/3, pp.202–239.
81. Ganesh, M. (2004), "Six Sigma: time for a reality check", Customer Management, Vol. 12, No. 2, pp. 38-39.
82. Garg, N. (2010) 'Major schemes of ministry of MSMES/ DC MSME for benefits of MSMEs', Indian Foundry Journal, Vol. 56, No. 11, pp.91–94.
83. Gibbons, P. (2006), "Improving overall equipment efficiency using a lean Six Sigma approach", International Journal of Six Sigma and Competitive Advantage, Vol. 2, No. 2, pp. 207-32.
84. Gijo, E. and Rao, T. (2005), "Six Sigma implementation – hurdles and more hurdles", Total Quality Management and Business Excellence, Vol. 16, No. 6, pp. 721-25.
85. Gijo, E.V. and Scaria, J. (2010) 'Reducing rejection and rework by application of Six Sigma methodology in manufacturing process', International Journal of Six Sigma and Competitive Advantage, Vol. 6, Nos. 1/2, pp.77–90.
86. Goh, T.-N. and Xie, M. (2004), "Improving on the Six Sigma paradigm", The TQM Magazine, Vol. 16, No. 4, pp. 235-40.
87. Goh, T.-N., Low, P., Tsui, K. and Xie, M. (2003), "Impact of Six Sigma implementation on stock price performance", Total Quality Management and Business Excellence, Vol. 14, No. 7, pp. 753-63.

88. Goh, T.-N., Tang, L.-C., Lam, S.-W. and Gao, Y.-F. (2006), "Six Sigma: a SWOT analysis", International Journal of Six Sigma and Competitive Advantage, Vol. 2, No. 3, pp. 233-42.
89. Gowen, C. III and Tallon, W. (2005), "Effect of technological intensity on the relationship among Six Sigma design, electronic business, and competitive advantage: a dynamic capability model", Journal of High Technology Management Research, Vol. 16, pp. 59-87.
90. Graves, S. (2002), "Six Sigma rolled through-put yield", Quality Engineering, Vol. 14, No. 2, pp. 257-66.
91. Green, F. (2006), "Six Sigma and the green belt perspective: a study in five companies", International Journal of Six Sigma and Competitive Advantage, Vol. 2, No. 3, pp. 291-300.
92. Hagemeyer, C., Gershenson, J. and Johnson, D. (2006), "Classification and application of problem solving quality tools: a manufacturing case study", The TQM Magazine, Vol. 18, No. 5, pp. 455-83.
93. Hahn, G. (2005), "Six Sigma: 20 key lessons learned", Quality and Reliability Engineering International, Vol. 21, No. 3, pp. 225-33.
94. Hahn, G., Hill, W., Hoerl, R. and Zinkgraf, S. (1999), "The impact of Six Sigma improvement: a glimpse into the future of statistics", The American Statistician, Vol. 53, No. 3, pp. 208-15.
95. Haikonen, A., Savolainen, T. and Ja¨rvinen, P. (2004), "Exploring Six Sigma and CI capability development: preliminary case study findings on management role", Journal of Manufacturing Technology Management, Vol. 15, No. 4, pp. 369-78.
96. Hammer, M. (2002), "Process management and the future of Six Sigma", IEEE Engineering Management Review, Vol. 30, No. 4, pp. 56-63.
97. Hamza, S. (2008), "Design process improvement through the DMAIC Six Sigma approach: a case study from the Middle East", International Journal of Six Sigma and Competitive Advantage, Vol. 4, No. 1, pp. 35-47.
98. Hare, L. (2005), "Linking statistical thinking to Six Sigma", International Journal of Six Sigma and Competitive Advantage, Vol. 1, No. 4, pp. 389-402.

99. Harrington, H. and Trusko, B. (2005), "Six Sigma: an aspirin for healthcare", International Journal of Health Care Quality Assurance, Vol. 18, Nos 6/7, pp. 487-515.
100. Harry, M.J. (1998), ''Six Sigma: a breakthrough strategy for profitability'', Quality Progress, May, pp. 60-64.
101. Henderson, K. and Evans, J. (2000), "Successful implementation of Six Sigma: benchmarking General Electric Company", Benchmarking: An International Journal, Vol. 7, No. 4, pp. 260-81
102. Hendricks, C. and Kelbaugh, R. (1998), "Implementing Six Sigma at GE", Journal for Quality and Participation, Vol. 21, No. 4, pp. 48-53.
103. Hild, C., Sanders, D. and Cooper, T. (2000), "Six Sigma on continuous processes: how and why it differs", Quality Engineering, Vol. 13, No. 1, pp. 1-9.
104. Ho, S., Xie, M. and Goh, T. (2006), "Adopting Six Sigma in higher education: some issues and challenges", International Journal of Six Sigma and Competitive Advantage, Vol. 2, No. 4, pp. 335-52.
105. Ho, Y., Chang, O. and Wang, W. (2008), "An empirical study of key success factors for Six Sigma green belt projects at an Asian MRO company", Journal of Air Transport Management, Vol. 14, No. 5, pp. 263-69.
106. Hoerl, R. (1998), "Six Sigma and the future of the quality profession", IEEE Engineering Management Review, Vol. 26, No. 3, pp. 87-94.
107. Hoerl, R. (2004), "One perspective on the future of Six Sigma", International Journal of Six Sigma and Competitive Advantage, Vol. 1, No. 1, pp. 112-19.
108. Hoerl, R., Montgomery, D., Lawson, C., Molnau, W., Elias, R., Abraham, B., MacKay, J., Snee, R., Pyzdek, T., Hill, W., Breyfogle, F., Enck, D., Meadows, B. and Bailey, S. (2001), "Six Sigma black belts: what do they need to know?", Journal of Quality Technology, Vol. 33, No. 4, pp. 391-435.
109. Holtz, R. and Campbell, P. (2004), "Six Sigma: its implementation in Ford's facility management and maintenance functions", Journal of Facilities Management, Vol. 2, No. 4, pp. 320-29.
110. Hong, G. and Goh, T. (2003), "Six Sigma in software quality", The TQM Magazine, Vol. 15, No. 6, pp. 364-73.

111. Hong, G. and Goh, T. (2004), "A comparison of Six Sigma and GQM approaches in software development", International Journal of Six Sigma and Competitive Advantage, Vol. 1, No. 1, pp. 65-75.
112. Horacio Soriano-Meier, Paul L. Forrester, Sibi Markose, Jose Arturo Garza-Reyes, (2011) "The role of the physical layout in the implementation of lean management initiatives", International Journal of Lean Six Sigma, Vol. 2, No. 3, pp. 254 – 69.
113. Hsu, Y., Pearn, W. and Wu, P. (2008), "Capability adjustment for gamma processes with mean shift consideration in implementing Six Sigma program", European Journal of Operational Research, Vol. 191, No. 2, pp. 516-28.
114. Hu, M. and Pieprzak, J. (2005), "Using axiomatic design to improve conceptual design robustness in design for Six Sigma (DFSS) methodology", International Journal of Six Sigma and Competitive Advantage, Vol. 1, No. 3, pp. 245-62.
115. Huq, A. (2006), "Six-Sigma implementation through competency-based perspective (CBP)", Journal of Change Management, Vol. 6, No. 3, pp. 277-89.
116. Husband, S.G. (1997), Innovation in Advanced Professional Practice: Doctor of Technology (Report No 2), Faculty of Science and Technology, Deakin University, Geelong.
117. Hutchins, D. (2000), "The power of Six Sigma in practice", Measuring Business Excellence, Vol. 4, No. 2, pp. 26-33.
118. Hwang, Y.-D. (2006), "The practices of integrating manufacturing execution system and Six Sigma methodology", The International Journal of Advanced Manufacturing Technology, Vol. 30, Nos 7/8, pp. 761-68.
119. Immaneni, A., McCombs, A., Cheatham, G. and Andrews, R. (2007), "Capital One banks on Six Sigma for strategy execution and culture transformation", Global Business and Organizational Excellence, Vol. 26, No. 6, pp. 43-54.
120. Ingle, S. and Roe, W. (2001), "Six Sigma black belt implementation", The TQM Magazine, Vol. 13, No. 4, pp. 273-80.
121. Ingram, D. (2000), "Six Sigma and process validation strategies – part III", Journal of Validation Technology, Vol. 7, No. 1, pp. 67-75.

122. Irani, F.S. (2008), "Reactions to Different Levels of Personalization of Feedback: Moderating Effect of Individualism", 89th annual convention of the American, Psychological Association, Los angeles, pp. 94-110.
123. Irani, F.S. (2011), "Understanding And Quantifying The Impact Of Freeman And Medoff's; What Do Unions Do? A Quarter Of A Century Later", Journal of Business & Economics Research, Vol. 9, No. 9, pp. 13-28.
124. Isaacson, G. (2008), "Six Sigma tympanostomy tube insertion: achieving the highest safety levels during residency training", Otolaryngology – Head and Neck Surgery, Vol. 139, No. 3, pp. 353-57.
125. Jeffery, A. (2005), "Integrating organization development and Six Sigma: Six Sigma as a process improvement intervention in action research", Organization Development Journal, Vol. 23, No. 4, pp. 20-31.
126. Jin, M., Switzer, M. and Agirbas, G. (2008), "Six Sigma and lean in healthcare logistics centre design and operation: a case at North Mississippi Health Services", International Journal of Six Sigma and Competitive Advantage, Vol. 4, No. 3, pp. 270-88.
127. Johnson, A. (2002), "Six Sigma in R&D", Research Technology Management, Vol. 45, No. 2, pp. 12-16.
128. Johnson, A. and Swisher, B. (2003), "How Six Sigma improves R&D", Research Technology Management, Vol. 46, No. 2, pp. 12-15.
129. Johnston, A., Maguire, L. and McGinnity, T. (2008), "Disentangling causal relationships of a manufacturing process using genetic algorithms and Six Sigma techniques", International Journal of Production Research, Vol. 46, No. 22, pp. 6251-68.
130. Johnstone, P., Hendrickson, J., Dernbach, A., Secord, A., Parker, J., Favata, M. and Puckett, M. (2003), "Ancillary services in the health care industry: is Six Sigma reasonable?", Quality Management in Health Care, Vol. 12, No. 1, pp. 53-63.
131. Jolly, S.S. and Singh, B.J. (2014), "An approach to enhance availability of repairable systems: A case study of SPMs", International Journal of Quality and Reliability Management, Vol. 31, No. 9, pp. 1031-1051.

132. Joshi, R. and Singh, B.J. (2013), "Statistical Comparison of CFD Turbulent Models: A Case study with Centrifugal Slurry Pump", International Journal of Modeling in Operational Management (IJMOM) Vol.3, No. 3/4, pp.241-266.
133. Joshi, R., Jolly, S.S. and Singh, B.J. (2013), "Effect of Float Design on the Performance of Rotameter using Computational Fluid Mechanics", International Journal of Engineering and Management Research (IJEMR), Vol. 3, No. 6, pp. 15-20.
134. Juras, P., Martin, D. and Aldhizer, D. (2007), "Adapting Six Sigma to help tame the SOX 404 compliance beast", Strategic Finance, Vol. 88, No. 9, pp. 36-41
135. Kanji, G. (2008), "Reality check of Six Sigma for business excellence", Total Quality Management and Business Excellence, Vol. 19, No. 6, pp. 575-82.
136. Kaoru Ishikawa (1990), "Introduction to Quality Control", Publisher: Productivity Press, Japan.
137. Kasushik, P. (2010), "Six sigma: from concept to application across different Indian industries", Phd thesis, NIT, Kurukshetra, pp. 45-58.
138. Kaushik, P. and Khanduja, D. (2008), "DM make up water reduction in thermal power plants using Six Sigma DMAIC methodology", Journal of Scientific and Industrial Research, Vol. 67, No. 1, pp. 36-42.
139. Khan, Z.A. and Al-Darrab, I.A. (2010) 'Taguchi techniques-based study on the effect of mobile phone conversation on driver's reaction time', Int. J. Quality and Reliability Management, Vol. 27, No. 1, pp.63–77.
140. Khare, M. K. (2011), "Innovation Management in Companies and Organizations", International Seminar with collaboration of IIM Ahmedabad and Technical University Berlin. Ahmedabad, March, 14-18, pp 55-68.
141. Khare, V. K., Soni, V. and Wani, V. P. (2007), "Role of Technical Universities in successful implementation of Technical Incubation", Science Tech Entrepreneur Magazine, pp 55-67.
142. Kleasen, K. (2007), "Building human resources strategic planning, process and measurement capability: using Six Sigma as a foundation", Organization Development Journal, Vol. 25, No. 2, pp. 37-41.

143. Klefsjo, B., Bergquist, B. and Edgeman, R. (2006), "Six Sigma and total quality management: different day, same soup?", International Journal of Six Sigma and Competitive Advantage, Vol. 2, No. 2, pp. 162-78.
144. Klefsjo", B., Wiklund, H. and Edgeman, R. (2001), "Six Sigma seen as a methodology for total quality management", Measuring Business Excellence, Vol. 5, No. 1, pp. 31-35.
145. Knowles, G., Johnson, M. and Warwood, S. (2004), "Medicated sweet variability: a Six Sigma application at a UK food manufacturer", The TQM Magazine, Vol. 16, No. 4, pp. 284-92.
146. Kovach, J. (2007), "Designing efficient Six Sigma experiments for service process improvement projects", International Journal of Six Sigma and Competitive Advantage, Vol. 3, No. 1, pp. 72-90.
147. Kuei, C.-H. and Madu, C. (2003), "Customer-centric Six Sigma quality and reliability management", International Journal of Quality & Reliability Management, Vol. 20, No. 8, pp. 954-64.
148. Kumar, M. (2007), "Critical success factors and hurdles to Six Sigma implementation: the case of a UK manufacturing SME", International Journal of Six Sigma and Competitive Advantage, Vol. 3, No. 4, pp. 333-51.
149. Kumar, M., Antony, J., Antony, F. and Madu, C. (2007), "Winning customer loyalty in an automotive company through Six Sigma: a case study", Quality and Reliability Engineering International, Vol. 23, No. 7, pp. 849-66.
150. Kumar, S., Strandlund, E. and Thomas, D. (2008), "Improved service system design using Six Sigma DMAIC for a major US consumer electronics and appliance retailer", International Journal of Retail & Distribution Management, Vol. 36, No. 12, pp. 970-94.
151. Kumi, S. and Morrow, J. (2006), "Improving self-service the Six Sigma way at Newcastle University Library", Program: electronic library and information systems, Vol. 40, No. 2, pp. 123-36.
152. Kuruuzum, O. and Akyuz, G. (2009) 'Modelling the relationship between product quality and process characteristics in process industry: an application', Int. J. Productivity and Quality Management, Vol. 4, No. 4, pp.461–475.

153. Kwak, Y.H. and Anbari, F.T. (2006), "Benefits, obstacles and future of Six Sigma approach", Technovation, Vol. 26, pp. 708-15.
154. Lanyon, S. (2003), "At Raytheon Six Sigma works, too, to improve HR management processes", Journal of Organizational Excellence, Vol. 22, No. 4, pp. 29-42.
155. Laosirihongthong, T., Rahman, S. and Saykhun, K. (2006), "Critical success factors of Six-Sigma implementation", International Journal of Innovation and Technology Management, Vol. 3, No. 3, pp. 303-19.
156. Lee, K.-C. and Choi, B. (2006), "Six Sigma management activities and their influence on corporate competitiveness", Total Quality Management and Business Excellence, Vol. 17, No. 7, pp. 893-911.
157. Lee-Mortimer, A. (2006), "Six Sigma: effective handling of deep rooted quality problems", Assembly Automation, Vol. 26, No. 3, pp. 200-04.
158. Lin, J., Tien, S. and Hsu, C. (2008), "The adoption of Six Sigma methodology to close learning-doing gap", Journal of Statistics and Management Systems, Vol. 11, No. 1, pp. 49-64.
159. Linderman, K., Schroeder, R. and Choo, A. (2006), "Six Sigma: the role of goals in improvement teams", Journal of Operations Management, Vol. 24, pp. 779-90.
160. Linderman, K., Schroeder, R., Zaheer, S. and Choo, A. (2003), "Six Sigma: a goal-theoretic perspective", Journal of Operations Management, Vol. 21, pp. 193-203.
161. Lipscomb, B. and Lewis, A. (2004), "The principles of Six Sigma", Risk Management, Vol. 51, No. 2, pp. 30-34.
162. Liu, X., Wang, S., Qiu, J., Zhu, J., Guo, Y. and Lin, Z. (2008), "Robust optimization in HTS cable based on design for Six Sigma", IEEE Transactions on Magnetics, Vol. 44, No. 6, pp. 978-81.
163. Lloyd, D. II and Holsenbach, J. (2006), "The use of Six Sigma in health care operations: application and opportunity", Academy of Health Care Management Journal, Vol. 2, pp. 41-50.
164. Lochner, R.H. (1990), "Mater, J.E. Designing for Quality; Quality", White Plains, NY.

165. Lok, P., Rhodes, J., Diamond, A. and Bhatia, N. (2008), "The Six Sigma approach in performance management to improve safety culture at work", International Journal of Six Sigma and Competitive Advantage, Vol. 4, No. 2, pp. 151-71.
166. Mahanti, R. (2005), "Six Sigma for software", Software Quality Professional, Vol. 8, No. 1, pp. 12-26.
167. Mahanti, R. and Antony, J. (2005), "Confluence of Six Sigma, simulation and software development", Managerial Auditing Journal, Vol. 20, No. 7, pp. 739-62.
168. Maleyeff, J. and Kaminsky, F. (2002), "Six Sigma and introductory statistics education", Education þ Training, Vol. 44, No. 2, pp. 82-89.
169. Malhotra, N.K. (1999), Marketing Research, An Applied Orientation, 3rd ed., Prentice-Hall, Upper Saddle River, NJ, p. 538-42.
170. Man, J. (2002), "Six Sigma and lifelong learning", Work Study, Vol. 51, No. 4, pp. 197-201.
171. Manikandan, G., Kannan, S. and Jayabalan, V. (2008), "Six Sigma-inspired sampling plan design to minimise sample size for inspection", International Journal of Productivity & Quality Management, Vol. 3, No. 4, pp. 472-95.
172. Manual, D. (2006), "Six Sigma methodology: reducing defects in business processes", Filtration and Separation, Vol. 43, No. 1, pp. 34-36.
173. Markarian, J. (2004), "Six Sigma: quality processing through statistical analysis", Plastics, Additives and Compounding, Vol. 6, No. 4, pp. 28-31.
174. Marti, F. (2005), "Lean Six Sigma method in phase 1 clinical trials: a practical example", Quality Assurance Journal, Vol. 9, No. 1, pp. 35-39.
175. Mazzola, M., Gentili, E. and Aggogeri, F. (2007), "SCOR, lean and Six Sigma integration for a complete industrial improvement", International Journal of Manufacturing Research, Vol. 2, No. 2, pp. 188-97.
176. McAdam, R. and Lafferty, B. (2004), "A multilevel case study critique of Six Sigma: statistical control or strategic change?", International Journal of Operations & Production Management, Vol. 24, No. 5, pp. 530-49.
177. McAdam, R., Hazlett, S. and Henderson, J. (2005), "A critical review of Six Sigma: exploring the dichotomies", International Journal of Organizational Analysis, Vol. 3, No. 2, pp. 51-174.

178. McCarty, T. and Fisher, S. (2007), "Six Sigma: it is not what you think", Journal of Corporate Real Estate, Vol. 9, No. 3, pp. 187-96.
179. McClusky, R. (2006), "The rise, fall and revival of Six Sigma", Measuring Business Excellence, Vol. 4, No. 2, pp. 6-17.
180. Mehta, A., Armenakis, A.A., Mehta, N., & Irani, F. (2006a), "Challenges and Opportunities of Business Process Outsourcing in India", Journal of Labor Research, Vol. 27, No. 3, pp 323-38.
181. Mekki, K. (2006), "Robust design failure mode and effects analysis in designing for Six Sigma", International Journal of Product Development, Vol. 3, Nos 3/4, pp. 292-304.
182. Mitra, A. (2004), "Six Sigma education: a critical role for academia", The TQM Magazine, Vol. 16, No. 4, pp. 293-302.
183. Mohammad, R., Keivan, K., Reda, A. and Mick, J. (2009) 'Data modeling for effective data warehouse architecture and design', International Journal of Information and Decision Sciences, Vol. 1, No.3, pp.282–300.
184. Montgomery, D., Burdick, R., Lawson, C., Molnau, W., Zenzen, F., Jennings, C., Shah, H., Sebert, D., Bowser, M. and Holcomb, D. (2005), "A university-based Six Sigma program", Quality and Reliability Engineering International, Vol. 21, No. 3, pp. 243-48.
185. Moorman, D. (2005), "On the quest for Six Sigma", The American Journal of Surgery, Vol. 189, No. 3, pp. 253-58.
186. Morgan, S. and Cooper, C. (2004), "Shoulder work intensity with Six Sigma", Nursing Management, Vol. 35, No. 3, pp. 28-33.
187. Motwani, J., Kumar, A. and Antony, J. (2004), "A business process change framework for examining the implementation of Six Sigma: a case study of Dow Chemicals", The TQM Magazine, Vol. 16, No. 4, pp. 273-83.
188. Nikhil, M. (2008) "Successful knowledge management implementation in global software companies", Journal of Knowledge Management, Vol. 12, No. 2, pp.42 – 56.

189. Nikhil, M., Oswald, S., & Mehta, A. (2007), "Infosys Technologies Ltd.: Improving Organizational Knowledge Flows", Journal of Information Technology, Vol. 22, No. 3, pp. 456-464.
190. Nonaka, I. and von Krogh, G. (2009) 'Tacit knowledge and knowledge conversion: controversy and advancement in organizational knowledge creation theory', Organization Science, Vol. 20, No. 3, pp.635–652.
191. Nonthaleerak, P. and Hendry, L. (2006), "Six Sigma: literature review and key future research areas", International Journal of Six Sigma and Competitive Advantage, Vol. 2, No. 2, pp. 105-61.
192. Pan, J.-N. and Cheng, M.-Y. (2008), "An empirical study for exploring the relationship between balanced scorecard and Six Sigma programs", Asia Pacific Management Review, Vol. 13, No. 2, pp. 481-96.
193. Panchal, S. (2010) 'Indian foundries: past-present-future', Indian Foundry Journal, Vol. 56, No. 11, pp.79–82.
194. Pandey, A. (2007), "Strategically focused training in Six Sigma way: a case study", Journal of European Industrial Training, Vol. 31, No. 2, pp. 145-62.
195. Park, S.H. and Kim, K.H. (2000), "A study of Six Sigma for R&D part", Quality Revolution, Vol. 1 No. 1, Korean Society for Quality Management, pp. 51-65.
196. Patterson, A., Bonissone, P. and Pavese, M. (2005), "Six Sigma applied throughout the lifecycle of an automated decision system", Quality and Reliability Engineering International, Vol. 21, No. 3, pp. 275-92.
197. Pepper, M.P.J. and Spedding, T.A. (2010) 'The evolution of lean Six Sigma', International Journal of Quality & Reliability Management, Vol. 27, No. 2, pp.138–155.
198. Perry, L. and Barker, N. (2006), "Six Sigma in the service sector: a focus on non-normal data", International Journal of Six Sigma and Competitive Advantage, Vol. 2, No. 3, pp. 313-33.
199. Pfeifer, T., Reissiger, W. and Canales, C. (2004), "Integrating Six Sigma with quality management systems", The TQM Magazine, Vol. 16, No. 4, pp. 241-9.

200. Pheng, L. and Hui, M. (2004), "Implementing and applying Six Sigma in construction", Journal of Construction Engineering and Management, Vol. 130, No. 4, pp. 482-89.
201. Pojasek, P. (2003), "Lean, Six Sigma, and the systems approach: management initiatives for process improvement", Environmental Quality Management, Vol. 13, No. 2, pp. 85-92.
202. Prabhushankar, G., Devadasan, S., Shalij, P. and Thirunavukkarasu, V. (2008), "The origin, history and definition of Six Sigma: a literature review", International Journal of Six Sigma and Competitive Advantage, Vol. 4, No. 2, pp. 133-50.
203. Prabhushankar, G.V., Devadasan, S.R., Shalij, P.R. and Thirunavukkarasu, V. (2009) 'Design of innovative Six Sigma quality management systems', Int. J. Process Management and Benchmarking, Vol. 3, No. 1, pp.76–100.
204. Prasada, G.P. and Reddy, V.V. (2010) 'Process improvement using Six Sigma – a case study in small scale industry', International Journal of Six Sigma and Competitive Advantage, Vol. 6, Nos. 1/2, pp.1–11.
205. Raisinghani, M., Ette, H., Pierce, R., Cannon, G. and Daripaly, P. (2005), "Six Sigma: concepts, tools, and applications", Industrial Management & Data Systems, Vol. 105, No. 4, pp. 491-505.
206. Rajagopalan, R., Francis, M. and Sua´rez, W. (2004), "Developing novel catalysts with Six Sigma", Research Technology Management, Vol. 47, No. 1, pp. 13-16.
207. Rajamanoharan, I. and Collier, P. (2006), "Six Sigma implementation, organisational change and the impact on performance measurement systems", International Journal of Six Sigma and Competitive Advantage, Vol. 2, No. 1, pp. 48-68.
208. Rao, K. and Rao, K. (2007), "Higher management education: should Six Sigma be added to the curriculum?", International Journal of Six Sigma and Competitive Advantage, Vol. 3, No. 2, pp. 156-70.
209. Rasis, D., Gitlow, H. and Popovich, E. (2002), "A fictitious Six Sigma green belt case study. II", Quality Engineering, Vol. 15, No. 2, pp. 259-74.

210. Ribardo, C. and Allen, T. (2003), "An alternative desirability function for achieving Six Sigma quality", Quality and Reliability Engineering International, Vol. 19, No. 3, pp. 227-40.
211. Ricondo, I. and Viles, E. (2005), "Six Sigma and its link to TQM, BPR, lean and the learning organisation", International Journal of Six Sigma and Competitive Advantage, Vol. 1, No. 3, pp. 323-54.
212. Rylander, D. and Provost, T. (2006), "Improving the odds: combining Six Sigma and online market research for better customer service", SAM Advanced Management Journal, Vol. 71, No. 1, pp. 15-19.
213. Sachin, Singh, B.J. and Dhull, V. (2013), "Six Sigma: From Concept to Implementation", AMMM-2013 National Conference, Haryana.
214. Sadagopan, P., Devadasan, S. and Goyal, S. (2005), "Three Six Sigma transitions and organisational preparedness exercise today's imperatives for tomorrow's success", International Journal of Six Sigma and Competitive Advantage, Vol. 1, No. 2, pp. 134-50.
215. Sahoo, A., Tiwari, M. and Mileham, A. (2008), "Six Sigma-based approach to optimize radial forging operation variables", Journal of Materials Processing Technology, Vol. 202, Nos 1-3, pp. 125-36
216. Sanders, D. and Hild, C. (2000), "A discussion of strategies for Six Sigma implementation", Quality Engineering, Vol. 12, No. 3, pp. 303-09.
217. Sandhya, B. and Costas, T. (2010) 'Identifying buyers with similar seller rating models and using their opinions to choose sellers in electronic markets', International Journal of Information and Decision Sciences, Vol. 2, No. 1, pp.1–16.
218. Sarkar, B. (2007), "Capability enhancement of a metal casting processes in a small steel foundry through Six Sigma: a case study", International Journal of Six Sigma and Competitive Advantage, Vol. 3, No. 1, pp. 56-71.
219. Savolainen, T. and Haikonen, A. (2007), "Dynamics of organizational learning and continuous improvement in Six Sigma implementation", The TQM Magazine, Vol. 19, No. 1, pp. 6-17.

220. Schroeder, R.G., Linderman, K., Liedtke, C. and Choo, A. (2008), "Six Sigma: definition and underlying theory", Journal of Operations Management, Vol. 26, No. 4, pp. 536-54.
221. Sekhar, H. and Mahanti, R. (2006), "'Confluence of Six Sigma, simulation and environmental quality: an application in foundry industries", Management of Environmental Quality: An International Journal, Vol. 17, No. 2, pp. 170-83.
222. Senapati, N. (2004), "Six Sigma: myths and realities", International Journal of Quality & Reliability Management, Vol. 21, No. 6, pp. 683-90.
223. Shahabuddin, S. (2008), "Six Sigma: issues and problems", International Journal of Productivity & Quality Management, Vol. 3, No. 2, pp. 145-60.
224. Shanmugam, V. (2007), "Six Sigma cup: establishing ground rules for successful Six Sigma deployment", Total Quality Management and Business Excellence, Vol. 18, No. 1, pp. 77-82
225. Sharma, U. (2003), "Implementing lean principles with the Six Sigma advantage: how a battery company realized significant improvements", Journal of Organizational Excellence, Vol. 22, No. 3, pp. 43-52.
226. Siddiqui, M. and Yang, K. (2010) 'Effective data analysis methods for incomplete two-level factorial experiments', Int. J. Experimental Design and Process Optimisation, Vol. 1, No. 4, pp. 348–364.
227. Sinan, C., Nuri, A. and Tugrul, U. (2010) 'Users and information technology: analysis of task information fit model', International Journal of Information and Decision Sciences, Vol. 2, No. 4, pp. 401–26.
228. Singh B.J. (2011), "Set-up Time Reduction: Opportunities and Strategies for Die Casting Foundries", Lambert Academic Publishing House, Germany. ISBN No. 978-3845405209.
229. Singh, B.J. and Bakshi, Y. (2012), "Six Sigma for Sustainable Energy Management: A Case Study", International Journal of Science Technology & Management, Vol. 2, No. 2, pp. 60-72.
230. Singh, B.J. and Bakshi, Y. (2013), "Back Up Power Systems: Efficacy of Six Sigma in Backup Power Systems", Grin Publishing House, Germany. ISBN No. 978-3-656-50736-9.

231. Singh, B.J. and Bakshi, Y. (2014), "Re-optimizing the Backup Power Systems through Six Sigma: A Case Study of Diesel Genset", International Journal of Lean Six Sigma, (IJLSS) Emerald. Vol. 5, No. 2, pp. 168-192.
232. Singh, B.J. and Joshi, R. (2015), "A Six Sigma Approach for Academic Excellence through Sustainable Assessment Criteria: A Case of an Indian University", Benchmarking: An International Journal (BIJ) Emerald. **(In Press)**
233. Singh, B.J. and Khanduja, D. (2009), Set-Up Time Reduction for Higher Productivity in Indian Foundries, Indian Institute of Foundry Men (IIF) organized by Chandigarh Chapter at Ludhiana, pp. 51-57.
234. Singh, B.J. and Khanduja, D. (2010a), "DMAICT: A Roadmap to Quick Changeovers", International Journal of Six Sigma and Competitive Advantage, Vol.6, No.1/2, pp.31-52.
235. Singh, B.J. and Khanduja, D. (2010b), "SME Sector of Punjab (India): From Renaissance to Recession", International Journal of Indian Culture and Business Management, Vol. 3, No. 5, pp. 544-559.
236. Singh, B.J. and Khanduja, D. (2010c), "SMED: For Quick Changeovers in Foundry SMEs", International Journal of Productivity and Performance Management, Vol. 59, No. 1, pp.98-116.
237. Singh, B.J. and Khanduja, D. (2010d), "Synergy of Cross Functional Process Mapping and SMED for Quick Changeovers: A Case Study", International Journal of Science Technology & Management, Vol-2, No. 2, pp.107-116.
238. Singh, B.J. and Khanduja, D. (2010e), DMAIC(S): Incubates Core Competencies in Indian Foundry SMEs: An Empirical Study in State of Punjab, 2nd International Conference on Production and Industrial Engineering (CPIE-2010), NIT, Jalandhar, India, pp. 1443-53.
239. Singh, B.J. and Khanduja, D. (2010f), Exploring Set-up Management in Indian Foundries, Advances in Mechanical Engineering (AME-2010), Organized by B.B.S.B.C.E. Fatehgarh Sahib, Punjab, pp. 70-76.
240. Singh, B.J. and Khanduja, D. (2011a), "Enigma of Six Sigma for Foundry SMEs in India: A Case Study", International Journal of Engineering Management and Economics, Vol. 2, No. 1, pp. 81-105.

241. Singh, B.J. and Khanduja, D. (2011b), "Does Analysis Matter in Six Sigma?: A Case Study", International Journal of Data Analysis Techniques and Strategies, Vol. 3, No. 3, pp. 300-324.
242. Singh, B.J. and Khanduja, D. (2011c), "Introduce Quality Processes through DOE: A Case Study in Die Casting Foundry", International Journal of Productivity and Quality Management, Vol. 8, No. 4, pp. 373-397.
243. Singh, B.J. and Khanduja, D. (2011d), "Design for Set-ups: A Step towards Quick Changeovers in Foundries", International Journal of Sustainable Designs, Vol. 1, No. 4, pp. 402-422.
244. Singh, B.J. and Khanduja, D. (2011e), Enhancing Competitiveness of Foundry SMEs through Design for Changeover (DFC): A Case Study, International Conference on Emerging Trends in Mechanical Engineering (ICETME-2011), Thapar University, Patiala, India, pp. 507-515.
245. Singh, B.J. and Khanduja, D. (2012a), "Essentials of D-phase to Secure the Competitive Advantage through Six Sigma", International Journal of Business Excellence, Vol. 5, No. 1/2, pp. 31-51.
246. Singh, B.J. and Khanduja, D. (2012b), "Ambience of Six Sigma in Indian Foundry SMEs-An Empirical Investigation", International Journal of Six Sigma and Competitive Advantage, Vol. 7, No. 1, pp. 12-40.
247. Singh, B.J. and Khanduja, D. (2012c), "Risk Management in Complex Changeovers through CFMEA: An Empirical Investigation", International Journal of Industrial and System Engineering, Vol. 10, No. 4, pp. 470-494.
248. Singh, B.J. and Khanduja, D. (2012d), "Scope of Six Sigma in Indian Foundry Operations", International Journal of Services and Operation Management, Vol. 13, No.1, pp.65-97.
249. Singh, B.J. and Khanduja, D. (2012e), "Developing Operation Measurement Strategy during Six Sigma Implementation: A Foundry Case Study", International Journal of Advanced Operation Management, Vol.4, No. 4, pp. 323-349.
250. Singh, B.J. and Khanduja, D. (2013), "Leveraging Six Sigma Disciplines to Reduce Scrap in Indian Foundry SMEs", 26th SEAANZ Conference, Sydney (NSW), Australia, pp. 6-24. (Held in July-2013)

251. Singh, B.J. and Khanduja, D. (2014), "Perspectives of Control Phase to manage Six Sigma implements: A Foundry Case", International Journal of Business Excellence (IJBEX), Vol.7, No.1, pp. 88-111.
252. Singh, B.J. and Sodhi, H.S. (2014), "Parametric Optimization of CNC Turning for Al-7020 with RSM", International Journal of Operational Research (IJOR), Inderscience, Vol. 20, No. 2, pp. 180-206.
253. Singh, B.J. and Sodhi, H.S. (2014), "RSM: A Key to Optimize Machining: Multi-Response Optimization of CNC Turning with Al-7020 Alloy", Anchor Academic Publishing, Diplomica Verlag, Hamburg. ISBN No. 978-3954892099.
254. Singh, B.J., Bakshi, Y. and Chauhan (2010), "Workshop Technology", A Text Book for Engineering Students, Eagle Publishing House, Jalandhar, Punjab.
255. Singh, B.J., Bakshi, Y. and Kaushik, P. (2011), Role of Six Sigma in Engineering Institutes: A Case Study, 41st ISTE Convention, India, pp. 7-20.
256. Singh, B.J., Khanduja, D. and Jaglan, P. (2012), "Six Sigma for Sustainable Energy Management in Foundries: A Case Study", SOM-2012 Conference, IIT, Delhi (Held in Dec 2012).
257. Singh, B.J., Khanduja, D. and Singh, A. (2011), "Demystifying MSA: A Structured Approach for Indian Foundry SMEs", International Journal of Quality and Innovation, Vol. 1, No. 3, pp. 217-236.
258. Singh, S.S., Singh, B.J. and Khanduja, D. (2014), "Synthesizing TBI-relevance in India through Six Sigma Approach", International Journal of Entrepreneurship and Innovation Management (IJEIM), Inderscience Publishers. **(In Press)**
259. Singla, R. and Singh, B.J. (2014), "Doe: A Key to Optimize Friction Stir Welding of Dissimilar Aluminum Alloys", International Journal of Engineering Research and Applications (IJERA), Special Issue, pp. 71-76.
260. Smith, L. and Phadke, M. (2005), "Some thoughts about problem solving in a DMAIC framework", International Journal of Six Sigma and Competitive Advantage, Vol. 1, No. 2, pp. 151-66.
261. Snee, R. (2004), "Six-Sigma: the evolution of 100 years of business improvement methodology", International Journal of Six Sigma and Competitive Advantage, Vol. 1, No. 1, pp. 4-20.

262. Snee, R.D. (1993), "Creating Robust Work Processes"' Qual. Prog. Vol. 26, No. 2, pp. 37–41.
263. Snee, R.D. (2009) 'Get moo-ving', Six Sigma Forum Magazine, Vol. 1, No. 1, pp.1–30.
264. Sodhi, H.S., Singh, B.J. and Khanduja, D. (2012), "Behavior Study of Cutting Parameters on Material Removal Rate for a Non-Ferrous Material While Turning on a CNC Turning Center", IETET-2012, Kurukshetra (Held in Nov 2012).
265. Sokovic, M., Pavletic, D. and Fakin, S. (2005), "Application of Six Sigma methodology for process design", Journal of Materials Processing Technology, Vol. 162-163, pp. 777-83.
266. Stevenson, W. and Mergen, E. (2006), "Teaching Six Sigma concepts in a business school curriculum", Total Quality Management and Business Excellence, Vol. 17, No. 6, pp. 751-56.
267. Su, C-T., Chiang, T.-L. and Chang, C.-M. (2006), "Improving service quality by capitalising on an integrated lean Six Sigma methodology", International Journal of Six Sigma and Competitive Advantage, Vol. 2, No. 1, pp. 1-22.
268. Taner, M., Sezen, B. and Anthony, J. (2007), "An overview of Six Sigma applications in healthcare industry", International Journal of Health Care Quality Assurance, Vol. 20, No. 4, pp. 329-40.
269. Tang, L., Goh, T.-N., Lam, S. and Zhang, C. (2007), "Fortification of Six Sigma: expanding the DMAIC toolset", Quality and Reliability Engineering International, Vol. 23, No. 1, pp. 3-18
270. Taylor, A. and Taylor, M. (2009) 'Operations management research: contemporary themes, trends and potential future directions', Int. J. Operations and Production Management, Vol. 29, No. 12, pp.1316–40.
271. Thakkar, J., Deshmukh, S. and Kanda, A. (2006), "Implementing Six Sigma in service sector using AHP and Alderfer's motivational model: a case of educational services", International Journal of Six Sigma and Competitive Advantage, Vol. 2, No. 4, pp. 353-76.
272. Thawani, S. (2004), "Six Sigma – strategy for organisational excellence", Total Quality Management and Business Excellence, Vol. 15, Nos 5/6, pp. 655-64.

273. Thirunavukkarasu, V., Devadasan, S., Prabhushankar, G., Murugesh, R. and Senthilkumar, K. (2008), "Conceptualisation of total Six Sigma function deployment through literature snapshots", International Journal of Applied Management Science, Vol. 1, No. 1, pp. 97-122.
274. Thomas, A. and Barton, R. (2006), "Developing an SME-based Six Sigma strategy", Journal of Manufacturing Technology Management, Vol. 17, No. 4, pp. 417-34.
275. Tomkins, R. (1997), "GE beats expected 13% rise", Financial Times, October 10, p. 22-28.
276. Van Den Heuvel, J., Does, R. and De Koning, H. (2006), "Lean Six Sigma in a hospital", International Journal of Six Sigma and Competitive Advantage, Vol. 2, No. 4, pp. 377-88.
277. Velazquez, M.A., Claudio, D. and Ravindran, R. (2010) 'Experiments in multiple criteria selection problems with multiple decision makers', Int. J. Operational Research, Vol. 7, No. 4, pp.413–28.
278. Vote, D. and Huston, J. (2005), "Six Sigma approach to improve surgical site infections: a key variable", American Journal of Infection Control, Vol. 33, No. 5, pp. 167-69.
279. Wang, F., Du, T. and Li, E. (2004), "Applying Six Sigma to supplier development", Total Quality Management and Business Excellence, Vol. 15, Nos 9/10, pp. 1217-29.
280. Webb, M. (2003), " Introduction to Six Sigma for marketing and sales", Impact communication Inc., September edition, pp.01-01.
281. Weinstein, L., Castellano, J., Petrick, J. and Vokurka, R. (2008), "Integrating Six Sigma concepts in an MBA quality management class", Journal of Education for Business, Vol. 83, No. 4, pp. 128-233.
282. Wessel, G. and Burcher, P. (2004), "Six Sigma for small and medium-sized enterprises", The TQM Magazine, Vol. 16, No. 4, pp. 264-72.
283. Wiklund, H. and Wiklund, P. (2002), "Widening the Six Sigma concept: an approach to improve organizational learning", Total Quality Management and Business Excellence, Vol. 13, No. 2, pp. 233-39.

284. Williamsen, M.M. (2005), " Six Sigma safety", June Edition, Professional Safety, pp. 41-49. www.asse.org.
285. Woodall, T. (2001), "Six Sigma and service quality: Christian Gro¨nroos revisited", Journal of Marketing Management, Vol. 17, Nos 5/6, pp. 595-607.
286. Wright, J. and Basu, R. (2008), "Project management and Six Sigma: obtaining a fit", International Journal of Six Sigma and Competitive Advantage, Vol. 4, No. 1, pp. 81-94.
287. Wu, C. and Lin, C. (2009) 'Case study of knowledge creation facilitated by Six Sigma', Int. J. Quality and Reliability Management, Vol. 26, No. 9, pp.911–932.
288. Wyper, B. and Harrison, A. (2000), "Deployment of Six Sigma methodology in human resource function: a case study", Total Quality Management and Business Excellence, Vol. 11, Nos 4/6, pp. 720-27.
289. Yang, C. (2004), "An integrated model of TQM and GE-Six-Sigma", International Journal of Six Sigma and Competitive Advantage, Vol. 1, No. 1, pp. 97-111.
290. Yang, C. and Yeh, T.-M. (2007), "An integrated model of Hoshin management and Six Sigma in high-tech firms", Total Quality Management and Business Excellence, Vol. 18, No. 6, pp. 653-65.
291. Yang, K., Yeh, T., Pai, F. and Yang, C.-C. (2008), "The analysis of the implementation status of Six Sigma: an empirical study in Taiwan", International Journal of Six Sigma and Competitive Advantage, Vol. 4, No. 1, pp. 60-80.
292. Yeh, D. (2007), "A decision process for Li-Ion battery suppliers selection based on the two-tuple fuzzy linguistic representation model and the Six Sigma DMAIC processes", Asia Pacific Management Review, Vol. 12, No. 6, pp. 299-310.
293. Yeung, S. (2007), "Integrating ISO 9001:2000 and Six Sigma into organisational culture", International Journal of Six Sigma and Competitive Advantage, Vol. 3, No. 3, pp. 210-27.
294. Yilmaz, M. and Chatterjee, S. (2000), "Six Sigma beyond manufacturing: a concept for robust management", The Quality Management Journal, Vol. 7, No. 3, pp. 67-78.

295. Yusof, S.M. and Aspinwall, E. (1999), ''Critical success factors for successful implementation of TQM in SMEs'', Total Quality Management, Vol. 10, Nos. 4-5, pp. 803-09.
296. Zhan, W. (2008), "A Six Sigma approach for the robust design of motor speed control using modeling and simulation", International Journal of Six Sigma and Competitive Advantage, Vol. 4, No. 2, pp. 95-113.
297. Zu, X., Fredendall, L. and Douglas, T. (2008), "The evolving theory of quality management:the role of Six Sigma", Journal of Operations Management, Vol. 26, No. 5, pp. 630-50
298. Zu, X., Zhou, H., Zhu, X. and Yao, D. (2011) "Quality management in China: The effects of firm characteristics and cultural profile", International Journal of Quality & Reliability Management, Vol. 28, No. 8, pp. 32-45.

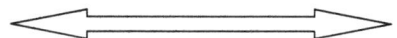

ANNEXURE-1

SIPOC DIAGRAM

Detailed SIPOC Diagram for Piston Foundry

Suppliers (Providers of the required resources)	Inputs (KPIVs) (Resources required by the process)	Process (Top level description of the activity)	Outputs (KPOVs) (Deliverables from the process)	Customers (Anyone who receives a deliverable from the process)
	1. AS PER SPECIFICATION	INSPECTION OF INCOMING MATERIALS	1. Al,Si,Cu,Ni,Mg Purity	
NALCO, FERROPEM	2. -DO-	- RAW MATERIALS	2. -DO-	
	3. -DO-	- ALLOY	3. -DO-	
	4. NIL	-CONVERTED ALLOY	4. ALLOY COMPOSITION (Control Plan)	
	5. PURITY OF GAS	-RETURNING MATERIAL	5. GAS FREE	
INOX IND LTD	6. MOISTURE	-ARGON/CL2/NITROGEN	6. GAS FREE	FOUNDRY SHOP
FOSECO	7. PHOSPHORUS CONTENT	-FLUXES	7. SILICON DISTRIBUTION	
	8. MIXING/DILUTION	-COPPER PHOSPHORUS	8. SHRINKAGE/DEPRESSION	
	9. SMOKE, FLAME, MESH SIZE	-DIE PAINT/TOOL PAINT	9. INCLUSION	
FOSECO	10. AS PER DRAWING	-FIBRE GLASS FILTER	10. IMPROPER CUTTING	
	11. AS PER DRAWING	-TCT CUTTER	11. BLANK AS PER DRAWING	
	12. DIMENSIONS	-DIE INSPECTION	12. CONFIRMED DIE	
			1. NO MIXUP OF INCOMING MATERIALS	
GENERAL STORE	1. SUITABLE SPACE	STORAGE INSIDE	2. IDENTIFICATION & CONSIGNMENT	FOUNDRY SHOP
			3. WISE STACKING OF MATERIALS	
OPERATOR	1. PROPER HANDLING	TRANSPORTATION TO FOUNDRY	1. MATERIAL REACHES SAFELY	FOUNDRY SHOP
	1. AS PER SPECIFICATION	ALLOY MAKING	1. ALLOY AS PER CONTROL PLAN OF FP/F/14	
	1.9% to 2.9%	copper	2.0% to 2.6%	
	14.8% to 18%	Silicon	15.5% to 17.5%	
GENERAL STORE	0.5% to 10% AS PER	Magnesium	0.7% to 1 %	CASTING SECTION
	0.35% to 0.8%	Nickle	0.5% to 0.7%	
	0.85% max	Iron	0.7% max	
	0.40% max	Maganse	0.3% max	
	20 PPM	Phosphorus	60 PPM	
	Balance	Aluminium	Balance	

	2. COMPOSITION AS PER SPECIF.	SPECTROMETER ARL L.C.0.1%	2. AS PER CONTROL PLAN OF FP/N/22	
PISTON LAB			TEMP 800±10 C	MELTING SECTION
MELTING SECTION	1. TEMPERATURE	TAPPING	1.TEMPERATURE	CASTING SECTION
	790±10 C		790±10 C	
OPERATOR	1. HANDLING WITHOUT SPLASH	TRANSPORTATION TO HOLDING FURNACE	1. MOLTEN METAL IN HOLDING FURNACE	CASTING SECTION
	QUANTITY = 300KG		QUANTITY OF H.F. = 350KG	
SHOP STORE	1. QUANTITY OF FLUX =300±50gms	FLUXING ADDITION OF COV-11,NUCLEANT,	1. GAS FREE	CASTING SECTION
	2. PRESSURE & FLOW OF GAS =2-4 kg/cm sq.	DEGASSING WITH CL2/ ARGON/N2,GAS	2. SILICON DISTRIBUTION	
	3. DE-GASSING TIME= 10±2 mins	REMOVE SLAG BY SPOON	3. ALLOY FREE FROM SLAG / DROSS	
	1. Density Index upto 0.5	SAMPLE INSPECTION FOR GAS	1. PROBE MOULD COMPLETELY FILLED	
	2. VAC tester L.C.=0.1	VAC TEST APPARATUS	2. DENSITY INDEX>1.2	
PISTON LAB	3. IMMERSION PYROMETER L.C. 1 DEGREE	MONITORING OF TEMPERATURE	3. DECISION AS PER MASTER FP/N/52	CASTING SECTION
		MICROSTRUCTURE CHECKING	4. RESULT OF BRICK MOULD TEST	
	3. SILICON DISTRIBUTION,GRAIN SIZE	MICROSCOPE L.C.100μ MAX.	5. SILICON DISTRIBUTION,GRAIN SIZE	
DIE REPAIR SECTION	1. DIE AS PER DRAWING	DIE INSPECTION,	DIE AS PER DRAWING	CASTING SECTION
		STAGE INSPECTION-II	1. AS PER SPECIFICATION OF IQ/F/05	
LINE INCHARGE	1. G.P.BOSS THICKNESS DIFFERENCE		AS PER SPECIFICATION OF IQ/F/04	
*	2.PIN HOLE DISPLACEMENT		1. G.P. BOSS THICKNESS	
	3.HEAD LOCATION		2. PIN HOLE DISPLACEMENT	
DIMENSIONAL CHECKER	4. DIE FITMENT, DIE & M/C BED CLEANING		3. HEAD PISTON	
	5. GUIDE BUSH FITMENT, OM CLEANING			
	6. CONDITION OF OM		4. INSURE PROPER MATING OF DIE	
	7. PIN STICK 0, BOTTOM SEATING	CASTING		
	8. WATER PRESSURE>4 Kg/cmsq.			CUTTING SECTION
OPERATOR	9. AIR PRESSURE>4 -5 Kg/cmsq.		1. CASTING AS PER SPECIFICATION OF IQ/F/03	
	10. POURING RATE, COOLING TIMER			
	11. METAL DENSITY INDEX<1.2			

CASTING MACHINE	1. CAST PISTON BLANK AS PER SAMPLE	STAGE INSPECTION -I	1. CAST PISTON BLANK CHECKED AS PER SAMPLE DISPLAYED & CORRECTIVE ACTION TAKEN IF REQUIRED.	CUTTING SECTION
	2. DIMENSIONAL CHECKS			
	3. VISUAL INSPECTIONS			
CASTING MACHINE	1. FINS/ BURR	RUNNER / RISER CUTTING	1. FILING OF FINS/BURR	HEAT TREATMENT SECTION
		SHEARING	2. CLAMPING PRESSURE	
			3. PROPER SHEARING FORCE	
CUTTING SECTION	1. AGING TEMPERATURE	AGE HARDENING	1. AS PER SPECIFICATION OF FP/N/24	FINAL INSPECTION
	2. HOLDING TIME		CHECKING OF HARDNESS AS PER IS : 2500 GIL-1 AQL 0.65 SAMPLING PLAN,	
	3. LOADING TIME NOT > 10MINS			
	4. UNLOADING TIME NOT > 10MINS		2. HARDNESS OF BLANK	
	5. WALL CLOCK L.C. 1 SEC.		3. CONTROL METHOD CHECKED BY FP/F/24	
	6. THERMOCOUPLE CALIBERATION.		4. OPERATOR'S SKILL	
HEAT TREATMENT	1. UNSOUND CASTINGS	STAGE INSPECTION -II REWORKING	1. VISUAL INSPECTION BLANKS AS PER SAMPLE DISPLAYED	BLANK STORE
HEAT TREATMENT	2. HARDNESS VARIATION	STAGE INSPECTION -II REWORKING	1. CHECKING OF HARDNESS IS : 2500 GIL-1 AQL 0.65 SAMPLING PLAN,OPERATOR'S SKILL. (IQ/F/09)	BLANK STORE
OPERATOR	1. BY FORK LIFTER	TRANSPORTATION OF BLANKS TO BLANK STORE	1. PROPER TRANSPORTATION OF BLANKS	MACHINE SHOP
			2. APPROPRIATE STACKING OF BLANKS	
VISUAL INSPECTOR	1. COLD LAP	VISUAL INSPECTION & STORAGE OF BLANKS	1. Proper storage of blanks	MACHINE SHOP
	2. HIT MARKS/DENT			
	3. FIN/BURR			
	4. BEND ON BLANKS			
	5. CUTTING ON BLANKS			
	6. BURR ON PIN HOLE			

ANNEXURE-2

DATA FOR Cpk CALCULATIONS (Before Improve Phase)

Shot No.	BT (mm)	Shot No.	BT (mm)
1	9.72	31	9.50
2	9.85	32	9.54
3	9.90	33	9.65
4	9.65	34	9.74
5	9.55	35	9.72
6	9.82	36	9.75
7	9.90	37	9.82
8	9.80	38	9.94
9	9.50	39	9.55
10	9.90	40	9.57
11	9.77	41	9.74
12	9.79	42	9.94
13	9.65	43	9.31
14	9.56	44	9.57
15	9.70	45	9.74
16	9.55	46	9.31
17	9.50	47	9.57
18	9.45	48	9.74
19	9.41	49	9.31
20	9.38	50	9.94
21	9.43	51	9.94
22	9.35	52	9.31
23	9.30	53	9.57
24	9.31	54	9.74
25	9.28	55	9.49
26	9.29	56	9.94
27	9.30	57	9.31
28	9.31	58	9.48
29	9.39	59	9.57
30	9.40	60	9.74

ANNEXURE-3

MULTI-REGRESSION DATA (Collected from Running Process)

Piston Number	Alloy Temperature	Stoppage (In Sec)	Die Temp	Results
1	755	103	125	SCRAP
22	755	65	145	SCRAP
45	755	61	190	SCRAP
50	755	79	225	SCRAP
35	755	25	256	GOOD
61	755	47	279	GOOD
14	755	26	281	GOOD
33	755	33	286	GOOD
7	755	26	298	GOOD
25	755	35	312	GOOD
80	757	7	315	GOOD
44	757	194	200	SCRAP
6	757	204	219	SCRAP
62	757	14	257	GOOD
60	757	6	268	GOOD
56	760	90	221	SCRAP
13	760	125	177	SCRAP
36	760	5	204	GOOD
12	763	2	300	GOOD
46	763	5	257	GOOD
72	763	3	373	GOOD
49	764	2	279	GOOD
59	764	2	294	GOOD
15	764	30	308	GOOD
34	764	12	319	GOOD
19	764	2	332	GOOD
26	764	146	241	SCRAP
52	764	97	231	SCRAP
3	764	3	253	GOOD
29	764	49	256	GOOD
65	764	10	269	GOOD
63	764	35	254	GOOD
42	764	64	247	GOOD
71	764	3	274	GOOD
74	765	10	301	GOOD
30	765	16	321	GOOD
79	760	4	269	GOOD
51	760	16	278	GOOD
8	760	7	289	GOOD

43	760	110	221	SCRAP
23	760	3	261	GOOD
32	763	4	255	GOOD
73	763	10	230	GOOD
75	763	5	220	GOOD
2	763	6	220	GOOD
58	765	40	298	GOOD
64	765	13	319	GOOD
28	765	10	334	GOOD
54	765	260	345	SCRAP
41	765	255	365	SCRAP
48	765	215	269	GOOD
66	752	52	279	GOOD
20	752	55	303	GOOD
31	752	99	313	GOOD
5	752	225	222	SCRAP
78	752	255	219	SCRAP
17	752	30	253	GOOD
55	752	45	278	GOOD
76	765	3	336	GOOD
39	765	10	351	GOOD
53	765	330	231	SCRAP
37	765	298	190	SCRAP
11	765	250	156	SCRAP
47	765	201	201	SCRAP
24	765	19	246	GOOD
4	765	60	264	GOOD
16	765	50	284	GOOD
38	752	44	298	GOOD
77	752	31	321	GOOD
10	752	64	281	GOOD
67	754	15	263	GOOD
57	754	2	298	GOOD
69	754	20	231	SCRAP
70	754	2	190	SCRAP
21	754	2	201	SCRAP
68	754	4	246	GOOD
40	754	2	276	GOOD
18	754	25	265	GOOD
9	754	18	258	GOOD
27	754	10	274	GOOD

ANNEXURE-4

TRAINING PERFORMA-1

Proposed Performa for Training of Engineers and Managers

IDENTIFICATION OF ANNUAL TRAINING NEEDS
(SUPERVISORS & MANAGERS)

DEPARTMENT : Piston Foundry

S.NO.	FUNCTION	EMPLOYEE NAME	EIN NO.	FUNCTIONAL TRAINING	BEHAVIORAL TRAINING
1	Manager Production Export	DINESH GUPTA	51175	Lean manufacturing, New Foundry Techniques.	Team Building, Role of Managers & Supervisors in Safety
2	Shift Incharge	A.L. JAGOTRA	60257	Rejection reduction, New Foundry Techniques.	Leadership skill, Role of Managers & Supervisors in Safety
3	Manager Melting	YOGESH DHIR	51377	Energy efficiency of melting furnaces, New Foundry Techniques.	Team Building, Role of Managers & Supervisors in Safety
4	Shift Incharge	S. MAHAJAN	51494	Rejection reduction, New Foundry Techniques.	Managerial skill improvement, Role of Managers & Supervisors in Safety
5	Shift Incharge	D.P. SINGH	62627	Rejection reduction, New Foundry Techniques.	Good industrial relation, Role of Managers & Supervisors in Safety
6	Manager Production	BALDEV SINGH	56707	Rejection reduction, New Foundry Techniques.	Managerial skill improvement, Role of Managers & Supervisors in Safety
7	Manager Production	SHALINDER PANDAV	57229	Non ferrous foundry technology, New Foundry Techniques.	Leadership skill, Role of Managers & Supervisors in Safety
8	Manager Production Export	RAVINDER SOOD	51871	Manufacturing, New Foundry Techniques.	Job delegation skill, Role of Managers & Supervisors in Safety
9	Shift Supervisor	GIRISH KUMAR	62852	SAP, Rejection reduction, New Foundry Techniques.	Managerial skill improvement, Role of Managers & Supervisors in Safety
10	Supervisor Cutting & Heat Treatment	P.C. GUPTA	51853	Waste Reduction, New Foundry Techniques.	Communication skill improvement, Role of Managers & Supervisors in Safety
11	Shift Supervisor	MAHESH INDER SINGH	62752	Productivity Improvement, SAP, New Foundry Techniques.	Leadership skill, Role of Managers & Supervisors in Safety
17	Shift Supervisor	MANPREET	52421	Rejection reduction, New Foundry Techniques.	Supervisory skill improvement, Role of Managers & Supervisors in Safety
18	Senior Manager	N.S.Bawa	2612	18001, EMS 14001, New Foundry Techniques.	Leadership Training
19	Senior Manager	Maninder Pal Singh	62252	Inventory Management, Foundry Process Controls	Leadership Training

DATE :

SIGNATURE
(HEAD OF DEPARTMENT)

ANNEXURE-5

TRAINING PERFORMA-2

Proposed Performa for Training Record of Workers

IDENTIFICATION OF ANNUAL TRAINING NEEDS
(WORKERS & STAFF)

DEPARTMENT: PISTON FOUNDRY

S.NO	E.I.NO	NAME	TRAINING NEEDS								
			Functional				Attitudinal			Safety	
			Introduction to six-sigma	Shop floor activities	Revised/TS 16949:2009	Quality Circle (Problem Solving Techniques)	5S- Better House Keeping	Team Building & Culture Development	KAIZEN	SAFETY AWARNESS	OHSAS -18001 & EMS-14001
1	50545	SITA RAM	✓	✓		✓					
2	50647	SHIV KUMAR	✓				✓	✓			
3	50653	RAJ KUMAR	✓	✓					✓		✓
4	50668	JARNAIL SINGH	✓	✓	✓				✓		
5	50706	JARNAIL SINGH	✓					✓		✓	✓
6	50724	DILBER KHAN	✓	✓		✓					
7	50744	BHARAT SINGH		✓				✓		✓	✓
8	50748	RAM LAL	✓	✓	✓		✓				✓
9	50766	SIYA RAM	✓	✓						✓	✓
10	50782	KARNAIL SINGH	✓					✓			✓

ANNEXURE-6

MSA CARD

Proposed Card for MSA Record Handling																
REFERENCE CODE	Name of SME-XYZ		AMENDENT			ISSUE STATUS										
			No.	Date		No.	Date									
IR/N-11/04						01	15.3.10									
			MSA PLAN													
S. NO.	PARAMETER	INSTRUMENT/ MEASURING DEVICE	TYPE OF STUDY	PLANNED ON YEAR –												RECORD IN FORMAT
				JAN	FEB	MAR	APR	MAY	JUN	JUL	AUG	SEP	OCT	NOV	DEC	
				Prepared By								Approved By				
Page No.																
Signature																
Date																

ANNEXURE-7

FMEA TABLE

Failure Mode and Effect Analysis for Casting Process

Process Steps	Requirements	Potential Failure Mode	Potential Effect(s) of Failure	Severity	Potential Cause(s) / Mechanism(s) of Failure	Current Process Controls Prevention	Occurrence	Current Process Controls Detection	Detection	RPN
PROCESS NO. 120 CASTING	BT DIFF. ± 0.3	More then > 0.3 mm	Internal depth less pin hole offset bottom thickness more in finish piston.	6	Die fitment not ok	Proper Fitment Ensured at the start	4	Visually checked at the start of shift by operator, visually step on bottom.	6	144
		BOSS DISTANCE MORE THAN SPECIFIED	Connecting Rod loose & loose mounting on Fixture.	6	Riser Bush Loose Gap in Core tools and thick tool paint.	Riser Bush proper Gap in core tools rectified if any and tools cleaned at the end of shift.	3	BT checked at the Checked at every new set up.	5	90
							3		6	108
PROCESS NO. 120 CASTING	DIMENSIONALLY ACURATE CASTING	BOSS DISTANCE LESS THAN SPECIFIED	Problem in connecting Rod Fitment & loose mounting on Fixture.	6	Tool wear out	Tool wear out	3	Checked at every new set up.	6	108
PROCESS NO. 120 CASTING	BOTTOM FITMENT OK/NO DAMAGE OF BOTTOM	VALVE SEAT DEPTH LESS OR MORE	Lesser Valve if damaged hence reduced life of product.	6	High spot on GP	Depression on GP	3	Boss distance	5	90
				6	Bottom Fitment/Cleaning not ok and Bottom	Proper fitment /cleaning ensured	4	First Piece check at the start of Die & checking by operator	6	144
PROCESS NO. 120 CASTING	CASTING FREE FROM COLD LAP DIE TEMP. < 180 C	COLD LAP ON BLANK	Reduced life of piston	6	Die is Cold	Preheating of Die with LPG & With molten metal at the start of shift) and	4	Visually checked by caster on blank,no cold lap on casting.	6	144
		COLD LAP BLANK	Reduced product life.	6	Mixing of warmers	Warmers are kept	4	Red Bit Provided	6	144
PROCESS NO. 120 CASTING	METAL TEMP. <750 STEADY POURING OF METAL IN THE DIE	Pin holes ON BLANK	Reduced product life.	7	Less pouring Slow pouring of metal in Die	Temperature Pouring is done with constant stream	4	Temperature checked Visually.	6	168
							4		6	168
PROCESS NO. 120 CASTING	LUFF STICKS OPEN	Pin holes	Reduced product life.	7	Air Entrapment	Luff Sticks	5	Visually	5	175
		Depression	Reduced product life.	7	Luff Sticks /	Weekly or as	5	Visually	5	175
	CASTING FREE FROM POROSITY DENSITY INDEX >1.2 TEMP. > 800 C	POROSITY	Reduction in Piston Life. Reduction in Piston life.	7	Gas Pick up from Casting temperature	Degassing done as Temperature	4	Sample Probe Temperature checked	3	84
									5	120
	CASTING SPOON PROPERLY COATED	METAL/INCLUSIONS STICKING WITH SPOON	Reduction in Piston life/ Spoon Life.	6	Poor Condition and cleaning of casting	Casting spoon Properly coated with Die Dress MS	3	Visually and coated at the end of Shift.	5	90
PROCESS NO. 120 CASTING	CASTING FREE FROM BLOW HOLES	BLOW HOLE IN CASTING	Reduction in life of	6	Air flow lines	Adequate air lines provided and cleaned when required.	3	Visually, outer mould cleaned at the start of every shaft	5	90
	NO PIN STICK AIR LEAKAGE	BLOW HOLES IN PIN	Reduction in life of	6	Pin Stick Air	Checked &	3	Visually	5	90
PROCESS NO. 120 CASTING	CASTING FREE FROM BLOW HOLES	BLOW HOLE IN CROWN	Reduction in life of	6	Cold Die	Preheating of die	3	Visually checked by	4	72
		Blow Hole		6	Melt Temp. High Sensor not working	Temperature Sensor working	4	Temperature checked Visually, myc tilting.	5	120
							4		4	96
PROCESS NO. 120 CASTING	CASTING FREE FROM SHRINKAGE	SHRINKAGE DEFECT	Reduction in life of	6	Metal temperature Metal temperature Less Metal /	Temperature Temperature Standard Die	4	Temperature checked Temperature checked Visually	5	120
				6			4		5	120
				6			4		5	120
PROCESS NO. 120 CASTING	DIE TEMP. >180 °C	SHRINKAGE DEFECT	Reduction in life of	6	Die Temperature	5 Warmers made	4	Visually	6	144

ANNEXURE-8

DATA FOR Cpk CALCULATIONS (After Improve Phase)

Shot no.	BT (mm)	Shot no.	BT (mm)
1	9.72	31	9.75
2	9.75	32	9.66
3	9.82	33	9.71
4	9.65	34	9.7
5	9.66	35	9.72
6	9.69	36	9.7
7	9.68	37	9.79
8	9.73	38	9.75
9	9.7	39	9.77
10	9.74	40	9.75
11	9.7	41	9.71
12	9.79	42	9.72
13	9.7	43	9.66
14	9.72	44	9.68
15	9.78	45	9.76
16	9.65	46	9.66
17	9.69	47	9.77
18	9.71	48	9.7
19	9.75	49	9.72
20	9.73	50	9.79
21	9.7	51	9.7
22	9.7	52	9.68
23	9.79	53	9.68
24	9.76	54	9.7
25	9.75	55	9.68
26	9.75	56	9.77
27	9.71	57	9.63
28	9.68	58	9.71
29	9.65	59	9.7
30	9.67	60	9.75

ANNEXURE-9

DATA FOR p-CHART (For the Month of December)

Day	Sample (n)	Scrap (np)	p-value (np/n)
1st Dec	425.00	41.65	0.10
2nd Dec	425.00	29.75	0.07
3rd Dec	425.00	46.75	0.11
4th Dec	425.00	59.50	0.14
5th Dec	425.00	76.50	0.18
6th Dec	425.00	78.63	0.19
7th Dec	425.00	76.50	0.18
8th Dec	425.00	68.00	0.16
9th Dec	425.00	59.50	0.14
10th Dec	425.00	42.50	0.10
11th Dec	425.00	42.50	0.10
12th Dec	425.00	42.50	0.10
13th Dec	425.00	34.00	0.08
14th Dec	425.00	38.25	0.09
15th Dec	425.00	42.50	0.10
16th Dec	425.00	42.00	0.10
17th Dec	425.00	42.00	0.10
18th Dec	425.00	41.00	0.10
19th Dec	425.00	39.00	0.09
20th Dec	425.00	38.00	0.09
21st Dec	425.00	41.00	0.10
22nd Dec	425.00	42.50	0.10
23rd Dec	425.00	42.50	0.10
24th Dec	425.00	29.75	0.07
25th Dec	425.00	34.00	0.08
26th Dec	425.00	34.00	0.08
27th Dec	425.00	38.25	0.09
28th Dec	425.00	38.25	0.09
29th Dec	425.00	29.75	0.07
30th Dec	425.00	34.00	0.08
31st Dec	425.00	29.75	0.07

ANNEXURE-10

EMPIRICAL ANALYSIS (Savings Incurred)

H-273 pistons: Already discussed in chapter 4 (refer table below for summary)

H-519 pistons:

Average scrap saved per month = 600 pistons (Approximately)

Scrap cost saved per piston = Rs 183/- (Approximately)

Total rejection cost saved per month = 600*183 = Rs 1,09,800/-

Total rejection cost saved per year = 1,09,800*12 = Rs 13,17,600/-

H-749 pistons:

Average scrap saved per month = 400 pistons (Approximately)

Scrap cost per piston = Rs 183/- (Approximately)

Total rejection cost saved per month = 400*183 = Rs 73,200/-

Total rejection cost saved per year = 73,200*12 = Rs 8,78,400/-

Net savings incurred are detailed as below:

Financial Parameters / Type of Pistons	H273	H519	H749
Scrap saved /Month (Nos.)	400	600	400
Scrap Cost/Piston (Rs.)	183	183	183
Rejection Cost saved/ Month (Rs.)	73,200	1,09,800	73,200
Rejection Cost saved /Year (Rs.)	8,78,400	13,17,600	8,78,400
Total Rejection Cost saved from H family Pistons Annually (Rs.)	30,74,400		

Total cost saved from scrap reduction of H-family pistons = 8,78,400 + 13,17,600 + 8,78,400

= Rs 30,74,400/- (Approximately)

FUTURE READINGS

1. A. Chakrabarty and K.C. Tan (2008), Case Study Analysis of Six Sigma in Singapore Service Organizations, IEEE 978-1-4244-1672-1
2. A.K. Sahoo, M.K. Tiwari, A.R. Mileham (2008), Six Sigma based approach to optimize radial forging operation variables, journal of materials processing technology, pp.125–136.
3. Agarwal, R. and Bajaj, N. (2008) 'Managing outsourcing process: applying Six Sigma', Business process management, Vol. 14, No. 6, pp. 829-837.
4. American Society for Quality, Glossary - Entry: Quality, retrieved 2008-07-20
5. Anbari, F.T., 2002. Six Sigma Method and Its Applications in Project Management, Proceedings of the Project Management Institute Annual Seminars and Symposium [CD], San Antonio, Texas. Oct 3–10. Project Management Institute, Newtown Square, PA.
6. Antony, J. (2008) 'Can Six Sigma be effectively implemented in SMEs?', International journal of productivity and performance management, Vol. 57, No. 5, pp. 420-423.
7. Arif, M., Verma, P.L., Manoria, A. and Bajpai, L. (2007) 'Six Sigma and theory of constraints for continuous improvement an integrated approach: A case study of spinning mill', Udyog pragati journal, Vol. 31, No. 1, pp. 35-42.
8. Babak Shirazi, Hamed Fazlollahtabar, Iraj Mahdavi (2010), A six sigma based multi-objective optimization for machine grouping control in flexible cellular manufacturing systems with guide-path flexibility, Advances in Engineering Software vol. 41, pp 865–873.
9. Bagaitkar, R. (2002) 'Making Six Sigma work-A case study of Tata honeywell ltd., Industry 2.0, September edition, pp. 30-35.
10. Bhote, K.R. and Bhote, A.K., World-Class Quality: Using Design of Experiments to Make it Happen, 2nd ed., 1991 (American Management Association: New York).
11. Bisgaard, S. and De Mast, J., After Six Sigma—what's next? Qual. Prog., 2006, 39(1), 30–36.

12. Bothe, D. R. (2002). Statistical reason for the 1.5s shift. Quality Engineering, 14(3): 479- 487.
13. Chen, H. C., & Yu, Y. W. (2008). Using a strategic approach to analysis the location selection for high-tech firms in Taiwan. Management Research News, 31(4), 228–244.
14. Chen, K.S., Huang, M.L. and Hung, Y.H., Process capability analysis chart with the application of CPM. Int. J. Prod. Res., 2008, 46(16), 4483–4499.
15. Cianfrani, Charles A.; West, John E. (2009). Cracking the Case of ISO 9001:2008 for Service: A Simple Guide to Implementing Quality Management to Service Organizations (2nd ed.). Milwaukee: American Society for Quality. pp. 5- "7. ISBN 978-0-87389-762-4.
16. Darshak A. Desai (2006), Improving customer delivery commitments the Six Sigma way: case study of an Indian small scale Industry, Int. J. Six Sigma and Competitive Advantage, Vol. 2, No. 1, 2006.
17. Das, P. (2005) 'Reduction in delay in procurement of material using Six Sigma philosophy', Total quality management, Vol. 16, No. 5, pp. 645-656.
18. Das, A., Pagell, M., Behm, M., Veltri, A. 2008. Toward a theory of the linkages between safety and quality. Journal of Operations Management, 26(4): 521-535.
19. Dasgupta, T., 2003. Using the Six Sigma metric to measure and improve the performance of a supply chain. Total Quality Management 14 (3), 355–366.
20. Desai, D.A. and Patil, M.B. (2006) 'Deploying Six Sigma in foundry industries for overall operational excellence', Proceeding of the international conference on global manufacturing and innovation, July 27-29, Coimbatore, India.
21. Ditahardiyani, P., Ratnayani, and Angwar, M. (2008) 'The quality improvement of primer packaging process using Six Sigma methodology, Journal technique industry, Vol. 10, No. 2, pp. 177-184.
22. Doran, C., "Using six sigma in the credit department," Credit Management. pp. 32-34, 2003.
23. Duncan, R.B., 1976. The ambidextrous organization: designing dual structures for innovation. In: Kilmann, R.H., Pondy, L.R., Slevin, D.P. (Eds.), The Management of Organization Design, vol. 1. North Holland, New York, pp. 167–188.

24. Ealey, Lance A. (1988). Quality by design: Taguchi methods and U.S. industry. Dearborn, Mich.: ASI Press. ISBN 978-1-55623-970-0. Cited by: Sriraman, Vedaraman, A primer on the Taguchi system of quality engineering, retrieved 2008-07-20.
25. Eckes, G., 2001. Making Six Sigma Last: Managing the Balance between Cultural and Technical Change. Wiley, New York.
26. Economic Survey (2001–2002) Government of India, Ministry of Finance, Economic Division.
27. Edgeman, R. L. (2000). Quoted in "New voices of quality: 21 for the 21st century". In: M. Maguire, (Ed.), Quality Progress (Vol. 33, pp. 31–39).
28. Elizabeth W. Kelly, Jonathan D. Kelly, Brian Hiestand, Kathy Wells-Kiser, Stephanie Starling, James W. Hoekstra (2010), Six Sigma Process Utilization in Reducing Door-to-Balloon Time at a Single Academic Tertiary Care Center, Progress in Cardiovascular Diseases, vol. 53, pp 219–226.
29. Evans, J.R., Lindsay, W.M., 2005. The Management and Control of Quality, sixth ed. South-Western, Mason, OH.
30. GE. 2002. General Electric Inc. annual report 2002: GE Inc.
31. Gopesh Anand, Peter T. Ward, Mohan V. Tatikonda (2010), Role of explicit and tacit knowledge in Six Sigma projects: An empirical examination of differential project success, Journal of Operations Management vol. 28, pp 303–315.
32. Hahn, G.J., Doganaksoy, N., Hoerl, R., 2000. The evolution of Six Sigma. Quality Engineering 12 (3), 317–326.
33. Haikonen, A., Savolainen, T. and Jarvinen, P. (2004) 'Exploring Six Sigma a CI capability development: preliminary case study findings on management role', Journal of manufacturing technology management, Vol. 15, No. 4, pp. 369-378.
34. Harry, M. and Schroeder, R. (2000) 'Six Sigma: The breakthrough management strategy revolutionizing the world top corporations', Double Day- a division of random house publication, First edition, February 2000.
35. Harry, M. J., Schroeder, R. (1999). Six Sigma: The Breakthrough Management Strategy Revolutionizing the World's Top Corporations. New York: Doubleday.

36. Harry, M., 1998. Six Sigma: a breakthrough strategy for profitability. Quality Progress 31 (5), 60–64.
37. Hashmi K., (2006) Introduction and Implementation of Total Quality Management, www.isixsigma.com
38. Hendericks, C., Kelbaugh, R., 1998. Implementing six sigma at GE. The Journal for Quality and Participation July/August.
39. Ho SL, Xie M, Goh TN (2006). Adopting Six Sigma in higher education: some issues and challenges. Int. J. Six Sigma and competitive advantage. 2: 335-352.
40. Hongbo Wang (2008), A Review of Six Sigma Approach: Methodology Implementation and Future Research, IEEE.
41. Hsu YC, Pearn WL, Pei CW (2008). Capability adjustment for Gamma processes with mean shift consideration in implementing six sigma program. Eur. J. Oper. Res., 191(2): 517-529.
42. J. Ravichandran (2006): Six-Sigma Milestone: An Overall Sigma Level of an Organization, Total Quality Management & Business Excellence, 17:8, 973-980.
43. Jiju, A. 2004. Some pros and cons of six sigma: an academic perspective. The TQM Magazine, 16(4): 303.
44. John Cox, Director of Training Parkland Memorial Hospital (2010), <www.iGrafx.com>
45. Kevin Linderman, Roger G. Schroeder, Srilata Zaheer, Adrian S. Choo (2003), Six Sigma: a goal-theoretic perspective, Journal of Operations Management, vol. 21, pp 193–203.
46. Kim, H., Kim, J., Baek, Y. and Moon, I. (2008) 'Teaching Six Sigma concepts in engineering college', American society for engineering education, AC2008-1028.
47. Klesjo, B. Wiklund, H. and Edgeman, R.L. (2001) 'Six Sigma seen as a methodology for total quality management', Measuring business excellence, Vol. 5, No. 1, pp. 31-35.
48. Kulkarni, P.R. (2002) 'Rehabilitation of sick small scale industry units', Productivity, Vol. 43, No.1, pp.123–132.

49. Kumi, S. and Morrow, J. (2006) 'Improving self service the Six Sigma way at Newcastle university library', Program: electronic library and information systems, Vol. 40, No. 2, pp.123–136.
50. Ladani, L.J., Das, D., Cartwright, J.L., Yenkner, R. and Razmi, J. (2006) 'Implementation of Six Sigma quality system in calestica with practical examples', International journal of Six Sigma and competitive advantage, Vol. 2, No.1, pp. 69-88.
51. Li, L., Markowski, E.P., Markowski, C. and Xu, L., Assessing the effects of manufacturing infrastructure preparation prior to enterprise information-systems implementation. Int. J. Prod. Res., 2008, 46(6), 1645–1665.
52. Linderman, K., Shroeder, R., Zaheer, S., Choo, A., 2003. Six Sigma: a goal theoretic perspective. Journal of Operations Management 21 (2), 193–203.
53. Linderman, K.W., Schroeder, R.G., Choo, A.S., 2006. Six Sigma: the role of goals in improvement teams. Journal of Operations Management 24 (6), 779–790.
54. Little, B. (2003) 'Six Sigma techniques improve the quality of e-learning', Industrial and commercial training, Vol. 35, No. 3, pp.104–108.
55. M. Sokovic, D. Pavletic, S. Fakin (2005), Application of Six Sigma methodology for process design, Journal of Materials Processing Technology, vol. 62–163, pp 777–783.
56. Madhani, P. M. (2012), Matching Compensation Strategies: Enhancing Competitiveness, SCMS Journal of Indian Management
57. Mahanti, R. and Antony, J. (2005) 'Confluence of Six Sigma, simulation and software development', Managerial auditing journal, Vol. 20, No. 7, pp.739–762.
58. Maleyeff, J. and Kaminsky, F.C. (2002) 'Six Sigma and introductory statistics education', Education + Training, Vol. 44, No. 2, pp.82–89.
59. Maleyeff, J. and Krayenvenger, D.E. (2004) 'Goal setting with Six Sigma mean shift determination', Aircraft engineering and aerospace technology, Vol. 76, No. 6, pp.577–583.
60. McCarthy, B. M. and R. Stauffer. 2001. Enhancing Six Sigma through Simulation with Igrafix Process for Six Sigma. In Proceedings of the 2001 Winter Simulation

Conference, ed. B. A. Peters, J. S. Smith, D. J. Medeiros, and M. W. Rohrer, 1241-1247. Piscataway, New Jersey: Institute of Electrical and Electronics Engineers, Inc.

61. McClusky, R., 2000. The Rise, fall, and revival of six sigma. Measuring Business Excellence 4 (2), 6–17.

62. McFadden, F.R., Six Sigma quality programs. Qual. Prog., 1993, 26(6), 37–42.

63. Motwani, J., Kumar, A. and Antony, J. (2004) 'A business process change framework for examining the implementation of Six Sigma: a case study of Dow chemicals', The TQM magazine, Vol. 16, No. 4, pp.273–283.

64. O'Neill, M. and Duvall, C. (2005) 'A Six Sigma quality approach to workplace evaluation', Journal of facilities management, Vol. 3, No. 3, pp.240–253.

65. P.O.C. et al (1991) "Increasing Productivity in Nigeria" Proceedings of the First National Conference on Productivity 1sty-3rd December 1987, National Productivity Centre, Macmillan, Nigeria. Pp. 98 -106.

66. Pande, P.S., Neuman, R.P., Cavanagh, R., 2000. The Six Sigma Way: How GE, Motorola, and Other Top Companies are Honing their Perfor- mance. McGraw-Hill, New York, NY.

67. Park, S.H. (2002) 'Six Sigma for productivity improvement: korean business corporation', Productivity journal, Vol. 43, No. 2, pp. 173-187.

68. Paul H. Selden (December 1998). "Sales Process Engineering: An Emerging Quality Application". Quality Progress: 59–63.

69. Pavel Mach, Jessica Guhqueta (2001), Utilization of the Seven Ishikawa Tools (Old Tools) in the Six Sigma Strategy, 24th International Spring Seminar on Electronics Technology.

70. Plotkin, H. (1999) 'Six Sigma: What it is and how to use it', Harvard management update, June edition, pp. 06-07.

71. Prabhakar Kaushik, Dinesh Khanduja (2010), Utilizing six sigma for improving pass percentage of students: A technical institute case study, Educational Research and Review Vol. 5 (9), pp. 471-483.

72. Prasad, J. (2002) 'Six Sigma in bulb manufacturing', Productivity journal, Vol. 43, No. 2, pp. 192-195.

73. Pyzdek T., 2003. The Six Sigma Handbook: A Complete Guide for Green Belts, Black Belts, and Managers at All Levels. McGraw-Hill, New York, NY.
74. Pyzdek T., Why Six Sigma is not TQM, <www.qualityamerica.com>, 2006.
75. Raju, R. (2000) 'Reducing axle rejection using problem solving tools-A case study', Industrial engineering journal, Vol.29, No.2, pp. 24-29.
76. Roberts, C.M., "Six sigma signals," Credit Union Magazine. Vol. 70 (1), pp. 40-43, 2004.
77. Roger G. Schroeder, Kevin Linderman, Charles Liedtke, Adrian S. Choo (2008), Six Sigma: Definition and underlying theory, Journal of Operations Management, vol. 26, pp 536–554.
78. Rohini. R, Dr. Mallikarjun .J (2011), Six Sigma: Improving the Quality of Operation Theatre, Procedia - Social and Behavioral Sciences, vol. 25, pp 273 – 280.
79. Rose, Kenneth H. (July, 2005). Project Quality Management: Why, What and How. Fort Lauderdale, Florida: J. Ross Publishing. p. 41. ISBN 1-932159-48-7.
80. Rowlands, H. and Antony, J. (2003) 'Application of design of experiments to a spot welding process', Assembly automation, Vol. 23, No. 3, pp. 273-279
81. S.J. Harjac, A. Atrens, C.J. Moss (2008), Six Sigma review of root causes of corrosion incidents in hot potassium carbonate acid gas removal plant, Engineering Failure Analysis vol.15, pp 480–496
82. Sanders, D., Hild, C.R., 2000. Six Sigma on business processes: common organizational issues. Quality Engineering 12 (4), 603–610.
83. Sarda, S.S. (2007) 'Linking QFD and Six Sigma for robust product design and process development' Industrial engineering journal, Vol. 36, No. 3, pp. 09-16.
84. Schroeder, R.G., Linderman, K., Liedtke, C., Choo, A.S., 2008. Six Sigma: definition and underlying theory. Journal of Operations Management 26, 536–554.
85. Sebastian Koziołek, Damian Derlukiewicz (2012) Method of assessing the quality of the design process of construction equipment with the use of DFSS (design for Six Sigma), Automation in Construction, vol. 22, pp 223–232.
86. Sharma O.P., Gupta V., Rathore G.S., Saini N.K., Sachdeva K. (2011), Six Sigma in Pharmaceutical industry and Regulatory Affairs: A Review, Journal of Natura Conscientia, Vol.2, Issue 1, January 2011

87. Sokovic, M., Pavletic, D. and Krulcic, E. (2006) 'Six Sigma process improvements in automotive parts production', Journal of achievements in materials and manufacturing engineering, Vol. 19, Issue 1, pp. 96-102.
88. Storey, D. (1994), Understanding the small business sector, UK: London, International Thomson Business Press.
89. Taguchi, G. (1992). Taguchi on Robust Technology Development. ASME Press. ISBN 978-99929-1-026-9.
90. Thomas, A. and Barton, R. (2006) 'Developing an SME based Six Sigma strategy', Journal of manufacturing technology management, Vol. 17, No. 4, pp.417–434.
91. U. Dinesh Kumar, David Nowicki, Jose´ Emmanuel Ramırez-Marquez, Dinesh Verma (2008), On the optimal selection of process alternatives in a Six Sigma implementation, Int. J. Production Economics vol. 111, pp 456–467.
92. Verma, R. (2005) 'Performance of small-scale industries (pre and post reform period)', Udyog Pragati, Vol. 29, No. 2, pp.35–41.
93. Wacker, J.G., 2004. A theory of formal conceptual definitions: developing theory-building measurement instruments. Journal of Operations Management 22 (6), 629–650.
94. Waxer, C. (2004) 'Is six sigma just for large companies? what about small companies?', www.sixsigma.com/library/content.
95. Weinberg, Gerald M. (1991). Quality Software Management: Volume 1. Systems Thinking. 1. New York, NY.: Dorset House. p. 7. ISBN 978-0-932633-72-9. OCLC 23870230.
96. Weiyong Zhang (2009), Six Sigma: A Retrospective and Prospective Study, POMS 20th Annual Conference Orlando, Florida U.S.A.
97. Westcott, Russell T. (2003). Stepping Up To ISO 9004: 2000 : A Practical Guide For Creating A World-class Oraganization. Paton Press. p. 17. ISBN 0-9713231-7-8.
98. Xingxing Zu, TinaL.Robbins, LawrenceD.Fredendall (2010), Mapping the critical links between organizational culture and TQM/Six Sigma practices, Int. J. Production Economics, vol. 123, pp 86–106.

99. Ying-Chin Ho, Ou-Chuan Chang, Wen-Bo Wang (2008), An empirical study of key success factors for Six Sigma Green Belt projects at an Asian MRO company, Journal of Air Transport Management 14 : 263–269
100. Yousef Amer, Lee Luong, William Y C Wang, Muhammad Azeem Ashraf, Zahid Qureshi (2007), Implementing Design for Six Sigma to Supply Chain Design.
101. Zhang, W., Xu, X. 2008. Six Sigma and information system (IS) project management: A revised theoretical model. Project Management Journal, 39(3): 59-74.
102. Hammer, M. and Goding, J. (2001) 'Putting Six Sigma in perspective', Quality magazine, October edition, pp. 58-62, www.qualitymag.com.
103. Hargrove, S.K. and Burge, L. (2002) 'Developing a Six Sigma methodology for improving retention in engineering education', 32nd ASEE/IEEE frontier in education conference, November 06-09, Boston, MA, pp. 20-24.
104. Harry, M. and Schroeder, R. (2000) 'Six Sigma: The breakthrough management strategy revolutionizing the worlds top corporations', Double Day- a division of random house publication, First edition, February 2000.
105. Hekmatpanah, M., Sadroddin, M., Shahbaz, S., Mokhtari, F. (2008) 'Six Sigma process and its impact on the organizational productivity', Proceedings of world academy of science, engineering and technology, Vol. 33, September, pp. 375-379.
106. Henderson, H. and Evans, J. (2000) 'Successful implementation of Six Sigma: benchmarking General Electric Company', Benchmarking: an international journal, Vol. 7, No 4, pp. 260-282.
107. Hensley, R.L. and Dobie, K. (2005) 'Assessing readiness for Six Sigma in a service setting', Managing service quality, Vol. 15, No. 1, pp.82–101.
108. Johnson, A. (2002) 'Six Sigma in R&D', Research technology management, March/April edition, pp. 12-16.

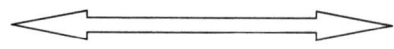

BIOGRAPHICAL NOTES

Dr. BIKRAM JIT SINGH is working as a Professor in Mechanical Engineering Department of Maharishi Markandeshwar University, Sadopur, Ambala (Haryana). He is PhD. from National Institute of Technology, Kurukshetra on Six Sigma and has done Masters of Engineering with distinction in Industrial Engineering from Thapar University, Patiala. He has a three year Professional Experience as a Manager in Light Alloy Foundry at FEDRAL MOGUL, INDIA (A US based Multi-National at Bahadurgarh, India.). He was undergone on the job training on SAP (PP-Module), QS-9000, ISO-TS 16949, SQC Tools and TPM for 2 years. He had worked as Lecturer in Mechanical Engineering Department for Two Years at RIMT-Institute of Engineering & Technology, Mandi Gobindgarh, India. He was also posted as HOD-Mechanical at BBSBPC, Fatehgarh Sahib, India for One Year. He has around 15 International Journal publications along with paper presented in number of International and National Conferences. He is a main and corresponding Author and can be contacted at chann461@yahoo.com .

Dr. DINESH KHANDUJA is a regular Professor from last 15 years at National Institute of Technology, Kurukshetra, India. Earlier he was posted as Assistant Professor in Regional Engineering College Kurukshetra. His fields of specializations are Industrial Engineering, Production Engineering and TQM. He had participated in more than 25 International and National Conferences and publishes around 38 Papers in various International and National Journals. He had also guided number of Master's and Doctorate students in the field of Six Sigma, Productivity Management, Waste Management and Entrepreneurship. He is also working as Head of Entrepreneurial Cell at NIT Kurukshetra India. He can be contacted at dineshkhanduja@yhaoo.com.